Fueling Innovation and Discovery

The Mathematical Sciences in the 21st Century

Committee on the Mathematical Sciences in 2025

Board on Mathematical Sciences and Their Applications

Division on Engineering and Physical Sciences

NATIONAL RESEARCH COUNCIL
OF THE NATIONAL ACADEMIES

THE NATIONAL ACADEMIES PRESS
Washington, D.C.
www.nap.edu

THE NATIONAL ACADEMIES
Advisers to the Nation on Science, Engineering, and Medicine

The **National Academy of Sciences** is a private, nonprofit, self-perpetuating society of distinguished scholars engaged in scientific and engineering research, dedicated to the furtherance of science and technology and to their use for the general welfare. Upon the authority of the charter granted to it by the Congress in 1863, the Academy has a mandate that requires it to advise the federal government on scientific and technical matters. Dr. Ralph J. Cicerone is president of the National Academy of Sciences.

The **National Academy of Engineering** was established in 1964, under the charter of the National Academy of Sciences, as a parallel organization of outstanding engineers. It is autonomous in its administration and in the selection of its members, sharing with the National Academy of Sciences the responsibility for advising the federal government. The National Academy of Engineering also sponsors engineering programs aimed at meeting national needs, encourages education and research, and recognizes the superior achievements of engineers. Dr. Charles M. Vest is president of the National Academy of Engineering.

The **Institute of Medicine** was established in 1970 by the National Academy of Sciences to secure the services of eminent members of appropriate professions in the examination of policy matters pertaining to the health of the public. The Institute acts under the responsibility given to the National Academy of Sciences by its congressional charter to be an adviser to the federal government and, upon its own initiative, to identify issues of medical care, research, and education. Dr. Harvey V. Fineberg is president of the Institute of Medicine.

The **National Research Council** was organized by the National Academy of Sciences in 1916 to associate the broad community of science and technology with the Academy's purposes of furthering knowledge and advising the federal government. Functioning in accordance with general policies determined by the Academy, the Council has become the principal operating agency of both the National Academy of Sciences and the National Academy of Engineering in providing services to the government, the public, and the scientific and engineering communities. The Council is administered jointly by both Academies and the Institute of Medicine. Dr. Ralph J. Cicerone and Dr. Charles M. Vest are chair and vice chair, respectively, of the National Research Council.

www.national-academies.org

COMMITTEE ON THE MATHEMATICAL SCIENCES IN 2025

THOMAS E. EVERHART, Chair, California Institute of Technology
MARK L. GREEN, Vice Chair, University of California, Los Angeles
TANYA STYBLO BEDER, SBCC Group, Inc.
JAMES O. BERGER, Duke University
LUIS A. CAFFARELLI, University of Texas at Austin
EMMANUEL J. CANDES, Stanford University
PHILLIP COLELLA, Lawrence Berkeley National Laboratory
DAVID EISENBUD, University of California, Berkeley
PETER WILCOX JONES, Yale University
JU-LEE KIM, Massachusetts Institute of Technology
YANN LeCUN, New York University
JUN LIU, Harvard University
JUAN MALDACENA, Institute for Advanced Study
JOHN W. MORGAN, Stony Brook University
YUVAL PERES, Microsoft Research
EVA TARDOS, Cornell University
MARGARET H. WRIGHT, New York University
JOE B. WYATT, Vanderbilt University

STAFF

SCOTT WEIDMAN, Director, Board on Mathematical Sciences and Their Applications
DANA MACKENZIE, Mathematics Writer
TOM ARRISON, Senior Staff Officer
MICHELLE SCHWALBE, Associate Program Officer
BARBARA WRIGHT, Administrative Assistant

BOARD ON MATHEMATICAL SCIENCES AND THEIR APPLICATIONS

C. DAVID LEVERMORE, Chair, University of Maryland
TANYA STYBLO BEDER, SBCC Group, Inc.
PATRICIA FLATLEY BRENNAN, University of Wisconsin
GERALD G. BROWN, Naval Postgraduate School
L. ANTHONY COX, JR., Cox Associates, Inc.
BRENDA DIETRICH, IBM Thomas J. Watson Research Center
CONSTANTINE GATSONIS, Brown University
DARRYLL HENDRICKS, UBS Investment Bank
KENNETH L. JUDD, Hoover Institution
DAVID MAIER, Portland State University
JAMES C. McWILLIAMS, University of California, Los Angeles
JUAN C. MEZA, University of California, Merced
JOHN W. MORGAN, Stony Brook University
VIJAYAN N. NAIR, University of Michigan
CLAUDIA NEUHAUSER, University of Minnesota, Rochester
J. TINSLEY ODEN, University of Texas at Austin
DONALD G. SAARI, University of California, Irvine
J.B. SILVERS, Case Western Reserve University
GEORGE SUGIHARA, University of California, San Diego
EVA TARDOS, Cornell University
KAREN L. VOGTMANN, Cornell University
BIN YU, University of California, Berkeley

STAFF

SCOTT WEIDMAN, Director
NEAL GLASSMAN, Senior Program Officer
MICHELLE SCHWALBE, Associate Program Officer
BARBARA WRIGHT, Administrative Assistant
BETH DOLAN, Financial Manager

Fueling Innovation and Discovery
The Mathematical Sciences in the 21st Century

Launched in 2010 with funding from the National Science Foundation, the National Academies' study *The Mathematical Sciences in 2025 (MathSci 2025)* is a forward-looking assessment of the current state of the mathematical sciences in the United States. The final report of the MathSci 2025 project will be released later in 2012. More information about the project is available at www.nas.edu/mathsci2025.

This publication, *Fueling Innovation and Discovery: The Mathematical Sciences in the 21st Century,* is a separate product being released in advance of the final report. It is based on the committee's identification of recent advances in the mathematical sciences or advances enabled by mathematical sciences research, drawn from the committee's assessment of the vitality of the discipline. This report is geared toward general readers who would like to know more about ongoing advances in the mathematical sciences and how these advances are changing our understanding of the world, creating new technologies, and transforming industries.

In selecting the topics for this report, the committee aimed to cover a range of mathematical sciences subfields and areas of impact, choosing topics where information was accessible and where developments could be described in a few pages. While the committee believes that all the topics covered are important and interesting, this publication is not intended to be a comprehensive selection of the most important developments in the mathematical sciences.

The committee worked primarily with mathematics writer Dana Mackenzie to prepare this report. It greatly appreciates his insights and hard work. During late 2010 and 2011, appropriate topics were identified, experts consulted, drafts prepared and revised, and accompanying images compiled. This report contains no committee conclusions or recommendations.

This report has been reviewed in draft form by individuals chosen for their diverse perspectives and technical expertise, in accordance with procedures approved by the National Academies' Report Review Committee. The review of this report was overseen by Samuel Fuller, Analog Devices, Inc. The purpose of this independent review is to provide candid and critical comments that will assist the institution in making its published report as sound as possible and to ensure that the report meets institutional standards for quality and objectivity. The review comments and draft manuscript remain confidential to protect the integrity of the process.

We wish to thank the following individuals for their review of this report:

JOHN BRADY, California Institute of Technology,
JAMES CARLSON, Clay Mathematics Institute,
ANNA GILBERT, University of Michigan,
MARVIN GOLDSTEIN, NASA Glenn Research Center,
RONALD GRAHAM, University of California, San Diego,
JON KETTENRING, Telcordia Technologies, Inc. (retired),
AROGYASWAMI PAULRAJ, Stanford University,
STEVEN STROGATZ, Cornell University,
LARRY WASSERMAN, Carnegie Mellon University,
EUGENE WONG, University of California, Berkeley, and
BIN YU, University of California, Berkeley.

Although the reviewers listed above have provided many constructive comments and suggestions, they were not asked to endorse the content of the report, nor did they see the final draft before its release. Responsibility for the final content of this report rests entirely with the committee and the institution.

CONTENTS

1   Introduction

3   Compressed Sensing: Through the Kaleidoscope

7   Eigenvectors: From the Mathematical Sciences to . . . an IPO

11   Mathematical Simulations: When the Lab Isn't Big Enough

18   Mathematical Sciences Inside . . . Tsunamis

20   Bayesian Inference: Not an Enigma Anymore

24   Diffusion Tensor Imaging: A New View of the Brain

29   Fast Multipole Method: A Long-Term Payoff

34   Mathematical Sciences Inside . . . the Battlefield

36   Cellular Automata: Sublimely Complex

40   Graph Spectra: Sparsest Cuts in Minimum Time

44   Bioinformatics: Interpreting the Human Genome

49   Geometry and Physics: Endlessly Intertwined

53   Probability and Statistical Physics: Connecting Microscopic and Macroscopic

56   Mathematical Sciences Inside . . . Inventions

Introduction

The mathematical sciences are part of everyday life. Modern communication, transportation, science, engineering, technology, medicine, manufacturing, security, and finance all depend on the mathematical sciences, which consist of mathematics, statistics, operations research, and theoretical computer science. In addition, there are very mathematical people working in theoretical areas of most fields of science and engineering who also contribute to the mathematical sciences. There is a healthy continuum between research in the mathematical sciences, which may or may not be pursued with an application in mind, and the range of applications to which mathematical science advances contribute. To function well in a technologically advanced society, every educated person should be familiar with multiple aspects of the mathematical sciences.

Although the mathematical sciences are pervasive, they are often invoked without an explicit awareness of their presence. For example, in the everyday operation of making a cell phone call, the mathematical sciences are essential in every step: We enter numbers in the decimal system, which are converted into sequences of bits (zeros and ones); next comes conversion to an electromagnetic signal; after an available receiver is located, the signal is transmitted and (finally) converted into the sound of our voice. Wireless technology uses techniques called "error correcting codes," "linear and nonlinear filtering," "hypothesis testing," "spatial multiplexing," "statistical waveform or parameter estimation," and these are built on tools of the mathematical sciences, such as matrix analysis, linear algebra, algebra, random matrices, graphical models, and so on.

More generally, the mathematical sciences contribute to modern life whenever data must be analyzed or when computational modeling and simulation is used to enable design and analysis of systems or exploration of "what-if" scenarios. The emergence of truly massive data sets across most fields of science and engineering, and in business, government, and national security, increases the need for new tools from the mathematical sciences. Because the mathematical sciences are independent of a particular scientific context, they can facilitate the translation of advances from one discipline to another.

The mathematical sciences provide a language—numbers, symbols, graphs, and diagrams—for expressing ideas in everyday life as well as in science, engineering, medicine, business, and the arts. Mathematical symbols, which are more universal than Chinese, English, or Arabic, allow communication across communities with completely dissimilar spoken and written languages.

The stories told here describe a number of recent advances made possible by research in the mathematical sciences.

Compressed Sensing

Through the Kaleidoscope

In the last two decades, two separate revolutions have brought digital media out of the pre-Internet age. Both revolutions were deeply grounded in the mathematical sciences. One of them is now mature, and you benefit whenever you go to a movie with computer-generated animation. The other revolution has just begun but is already redefining the limits of feasibility in some areas of biological imaging, communication, remote sensing, and other fields of science.

The first could be called the "wavelet revolution." Wavelets are a mathematical method for isolating the most relevant pieces of information in an image or in a signal of any kind (acoustic, seismic, infrared, etc.). There are coarse wavelets for identifying general features and fine wavelets for identifying particular details. Prior to wavelets, information was represented in long, cumbersome strings of bits that did not distinguish importance.

The central idea of wavelets is that for most real-world images, we don't need all the details (bytes) in order to learn something useful. In a 10-megapixel image of a face, for instance, the vast majority of the pixels do not give us any useful information. The human eye sees the general features that connote a face—a nose, two eyes, a mouth—and then focuses on the places that convey the most information, which tend to be edges of features. We don't look at every hair in the eyebrow, but we do look at its overall shape. We don't look at every pixel in the skin, because most of the pixels will be very much like their neighbors. We do focus on a patch of pixels that contrast with their neighbors—which might be a freckle or a birthmark or an edge.

Now much of this information can be represented much more compactly as the overlapping of a set of wavelets, each with a different coefficient to capture its weight or importance. In any typical picture, the weighting amplitude of most of the wavelets will be near zero, reflecting the absence of features at that particular scale. If the model in the photograph doesn't have a blemish on a particular part of her skin, you won't need the wavelet that would capture such a blemish. Thus you can compress the image by ignoring all of the wavelets with small weighting coefficients and keeping only the others. Instead of storing 10 million pixels, you may only need to store 100,000 or a million coefficients. The picture reconstructed from those coefficients will be indistinguishable from the original to the human eye.

Curiously, wavelets were discovered and rediscovered more than a dozen times in the 20th century—for example, by physicists trying to localize waves in time and frequency and by geologists trying to interpret Earth movements from seismograms. In 1984, it was discovered that all of these disparate, ad hoc techniques for decomposing a signal into its most informative pieces were really the same. This is typical of the role of the mathematical sciences in science and engineering: Because they are independent of a particular scientific context, the mathematical sciences can bridge disciplines.

Once the mathematical foundation was laid, stronger versions of wavelets were developed and an explosion of applications occurred. Some computer images could be compressed more effectively. Fingerprints could be digitized. The process could also be reversed: Animated movie characters could be built up out of wavelets. A company called Pixar turned wavelets (plus some pretty good story ideas) into a whole series of blockbuster movies (see Figure 1).

In 2004, the central premise of the wavelet revolution was turned on its head with some simple questions: Why do we even bother acquiring 10 million pixels of information if, as is commonly the case, we are going to discard 90 percent or 99 percent of it with a compression algorithm? Why don't we acquire only the most relevant 1 percent of the information to start with? This realization helped to start a second revolution, called compressed sensing.

Answering these questions might appear almost impossible. After all, how can we know which 1 percent of information is the most relevant until we have acquired it all? A key insight came from the interesting application of how to reconstruct a magnetic resonance image (MRI) from insufficient data. MRI scanners are too slow to allow them to capture dynamic images (videos) at a decent resolution, and they are not ideal for imaging patients such as children, who are unable to hold still and might not be good candidates for sedation. These challenges led to the discovery that MRI test images could, under certain conditions, be reconstructed perfectly—not approximately, but perfectly—from a too-short scan by a mathematical method called L1 (read as "ell-one") minimization. Essentially, random measurements of the image are taken, with each measurement being a randomly weighted average of many randomly selected pixels. Imagine replacing your camera lens with a kaleidoscope. If you do this again and again, a million times, you can get a better image than you can from a camera that takes a 10-megapixel photo through a perfect lens.

After all, how can we know which 1 percent of information is the most relevant until we have acquired it all?

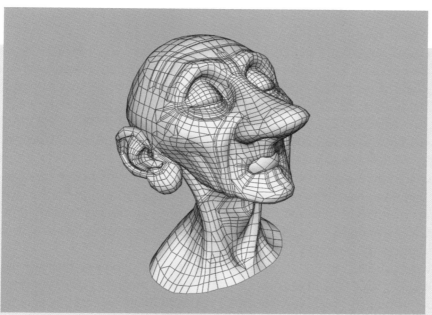

1 / *Stills from an animated short film called "Geri's Game," released by Pixar Animation Studios in 1997, which received an Academy Award in 1998. It was the first animated film to use subdivision surfaces, a mathematical technique based on wavelet compression. Wavelets allow computers to compress an image into a smaller data file. Subdivision surfaces do the reverse: They allow the computer to create a small data file that can be manipulated and then uncompressed to create lifelike images of something that never existed—in this case, an old man playing chess in the park. The top image shows the subdivision surface. The image below shows an actual frame from the movie. © 1997 Pixar.* /

The magic lies, of course, in the mathematical sciences. Even though there may be millions of scenes that would reproduce the million pictures you took with your kaleidoscopic camera, it is highly likely that there will be only one sparse scene that does. Therefore, if you know the scene you photographed is information-sparse (e.g., it contains a heart and a kidney and nothing else) and measurement noise is controlled, you can reconstruct it perfectly. L1 minimization happens to be a good technique for zeroing in on that one sparse solution. Compressed sensing actually built on, and helped make coherent, ideas that had been applied or developed in particular scientific contexts, such as geophysical imaging and theoretical computer science, and even in mathematics itself (e.g., geometric functional analysis). Lots of other reconstruction algorithms are possible, and a hot area for current research is to find the ones that work best when the scene is not quite so sparse.

As with wavelets, seeing is believing. Compressed sensing has the potential to cut down imaging time with an MRI from 2 minutes to 40 seconds. Other researchers have used compressed sensing in wireless sensor networks that monitor a patient's heartbeat without tethering him or her to an electrocardiograph. The sensors strap to the patient's limbs and transmit their measurements to a remote receiver. Because a heartbeat is information-sparse (it's flat most of the time, with a few spikes whose size and timing are the most important information), it can be reconstructed perfectly from the sensors' sporadic measurements.

Compressed sensing is already changing the way that scientists and engineers think about signal acquisition in areas ranging from analog-to-digital conversion to digital optics and seismology. For instance, the country's intelligence services have struggled with the problem of eavesdropping on enemy transmissions that hop from one frequency to another. When the frequency range is large, no analog-to-digital converter is fast enough to scan the full range in a reasonable time. However, compressed sensing ideas demonstrate that such signals can be acquired quickly enough to allow such scanning, and this has led to new analog-to-digital converter architectures.

Ironically, the one place where you aren't likely to find compressed sensing used, now or ever, is digital photography. The reason is that optical sensors are so cheap; they can be packed by the millions onto a computer chip. Even though this may be a waste of sensors, it costs essentially nothing. However, as soon as you start acquiring data at other wavelengths (such as radio or infrared) or in other forms (as in MRI scans), the savings in cost and time offered by compressed sensing take on much greater importance. Thus compressed sensing is likely to continue to be fertile ground for dialogue between mathematicians and all kinds of scientists and engineers.

> The magic lies, of course, in the mathematical sciences.

Eigenvectors
From the Mathematical Sciences to . . . an IPO

In 1997, when Sergey Brin and Larry Page were graduate students at Stanford, they wrote a short paper about an experimental search engine that they called Google. Brin and Page's idea—which was based on the research of many mathematical scientists—was to give each Web page a ranking, called PageRank, that indicates how authoritative it is. Your PageRank will improve if a lot of other Websites link to your Website. Intuitively, those other pages are casting a vote for your page. Also, Brin and Page assumed that a vote from a page that is itself quite authoritative should count for more. Thus your PageRank is a function of the PageRanks of all the pages that link to you.

The genius of Brin and Page's PageRank was its ability to harness human judgments without explicitly asking for them. Every link to a Web page is an implicit vote for the relevance of that page. In an exploding Internet, its simplicity also turned out to be of paramount importance. The calculation could be done offline, and thus it could be applied to the entire Web. PageRank represented a major advance over approaches to Internet search that were based on matching words or strings on a page. These earlier search engines returned far too many results, even in a drastically smaller Internet than today.

The PageRank algorithm seems to pose a chicken-and-egg paradox: To compute one PageRank, you already need to know all of the other PageRanks. However, Brin and Page recognized that this challenge is a form of a well-known type of math problem, known as the eigenvector problem. A vector (in this case) is just a list of numbers, such as the list of the PageRanks of all pages on the Web. If you apply the PageRank algorithm to a collection of vectors, most will be changed, but the true PageRank vector persists: It is not changed by the algorithm.

This kind of "persistent" vector is known in mathematics as an eigenvector (eigen being the German word for "characteristic"). Eigenvectors have appeared in numerous contexts over the centuries. The concept (though not the terminology) first arose in the work of the 18th century mathematician Leonhard Euler on the rotation of solid bodies. Because any rotation in space must have an axis—a line that persists in the same direction throughout the rotation—Euler recognized that the axis and angle of rotation characterize the rotation (which justifies the term "characteristic," at least in this context).

Fast forward a century or so, and you find eigenvectors used again in quantum physics. The motion of electrons is described by Schrödinger's equation, formulated in 1926 by Austrian physicist Erwin Schrödinger. They do not orbit atomic nuclei in circles or ellipses in the way that planets orbit the Sun. Instead, their orbits form complicated three-dimensional shapes that are determined by the eigenvectors of Schrödinger's equation. By counting the number of these solutions, you can tell how many electrons fit in each energy level or orbital of an atom, and in this way you can start to explain the patterns and periodicities of the periodic table.

Fast forward again to the present, and you can find the same concept used in genomics. Imagine that you have a large array of data; for example, the level of activity of 3,000 genes in a cell at 20 different times. Although the cell has thousands of genes, it does not have that many biologically meaningful processes. Some of the genes may work together to repel an invader. Other genes may be involved in cell division or metabolism. But the rest may not be doing much of anything, at least while you are watching them; their activity just amounts to random noise. The eigenvectors of the data set correspond to the most relevant patterns in the data, those which persist through the noise of chance variation. Figure 2 (on page 10) shows networks of genes found using eigenvectors. One eigenvector (the term "eigengene" has even been coined here) might correspond to genes that control metabolism. Another might consist of genes activated during cell division. The mathematics identifies the gene networks that appear most tied to biological activity, but it cannot tell what the networks do. That is up to the biologist.

Singular value decomposition (SVD) is a purely mathematical technique to pick out characteristic features in a giant array of data by finding eigenvectors. The idea is something like this: First you look for the one vector that most closely matches all of the rows of data in the array; that is the first eigenvector. Then you look for a second vector that most closely matches the residual variations after the first eigenvector has been subtracted out. This is the second eigenvector. The process can, of course, be repeated. For the PageRank example, only the first eigenvector is used. But in other applications, such as genomics, more than one eigenvector may be biologically significant.

Given the general applicability of eigenvector approaches, perhaps it is not too surprising that Google's PageRank—an algorithm that involves no actual understanding of your search query—could rank Web sites better than algorithms that attempted to analyze the semantic content of Web pages. However, the ability to use this eigenvector approach with real data that are random or contain much uncertainty is the key to PageRank. Within a few years, everybody was using Google, and "to google" had become a verb. When Brin and Page's company went public in 2004, its initial public stock offering raised $27 billion.

In many other applications, finding eigenvectors through SVD has proved to be effective for aggregating the collective wisdom of humans. From 2006 to 2009, another hot Internet company ran a competition that led to a number of advances in this field.

Netflix, a company that rents videos and streams media over the Internet, had developed a proprietary algorithm called Cinematch, which could predict the number of stars (out of five) a user would give a movie, based on the user's past ratings and the ratings of other users. However, its predictions were typically off by about 0.95 stars. Netflix wanted a better way to predict its customers' tastes, so in 2006 it offered a million-dollar prize for the first person or team who could develop an algorithm that would be 10 percent better (i.e., its average error would be less than about 0.85 stars). The company publicly released an anonymized database of 100 million past ratings by nearly half a million users so that competitors could test their algorithms on real data.

Within a few years, everybody was using Google, and "to google" had become a verb.

Rather unexpectedly, the most effective single method in the competition turned out to be good old-fashioned SVD. The idea is roughly as follows: Each customer has a specific set of features that they like in a movie—for instance, whether it is a drama or a comedy, whether it is a "chick flick" or a "guy flick," or who the lead actors are. A singular value decomposition of the database of past ratings can identify the features that matter most to Netflix customers. Just as in the genomics example, the mathematical sciences cannot say what the features are, but they can tell when two movies have the same constellation of factors. By combining a movie's scores for each feature with the weight that a customer assigns to those features, it can predict the rating the customer will give to the movie.

The team that won the Netflix Prize combined SVD with other methods to reach an improvement of just over 10 percent. Not only that, the competition showed that computer recommendations were better than the judgment of any human critic. In other words, the computer can predict how much your best friend will like a movie better than you can.

The above examples attest to the remarkable ability of eigenvector methods (often in combination with other techniques) to extract information from vast amounts of noisy data. Nevertheless, plenty of work remains to be done. One area of opportunity is to speed up the computation of eigenvectors. Recently mathematicians have found that "random projections" can compress the information in a large matrix into a smaller matrix while essentially preserving the same eigenvectors. The compressed matrix can be used as a proxy for the original matrix, and SVD can then proceed with less computational cost.

One of Google's biggest challenges is to guard the integrity of PageRanks against spammers. By building up artificial networks of links, spammers undercut the underlying assumption that a Web link represents a human judgment about the value of a Web page. While Google has refined the PageRank algorithm many times over to ferret out fake links, keeping ahead of the spammers is an ongoing mathematical science challenge.

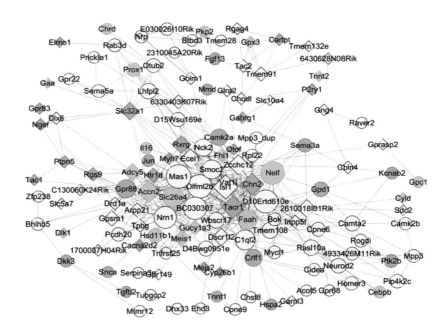

2 / *When the gene expressions for the C57BL/6J and A/J strains of mice are compared, it is possible to find gene networks using eigenvectors that are specific for brain regions, independent of genetic background. Image from S. de Jong, T.F. Fuller, E. Janson, E. Strengman, S. Horvath, M.J.H. Kas, and R.A. Ophoff, 2010, Gene expression profiling in C57BL/6J and A/J mouse inbred strains reveals gene networks specific for brain regions independent of genetic background,* **BMC Genomics 11:20.** * /*

Mathematical Simulations

When the Lab Isn't Big Enough

Computer simulations, which are built on mathematical modeling, are used daily in scientific research of all types, for informing decision making in business and government, including national defense, and for designing and controlling complex systems such as those for transportation, utilities, and supply chains, and so on. Simulations are used to gain insight into the expected quality and operation of those systems and to carry out what-if evaluations of systems that may not yet exist or are not amenable to experimentation.

As an example, one of the most important and spectacular events in the universe is the explosion of a star into a supernova. Such explosions seeded our own solar system with all of its heavier elements; they also have taught us, indirectly, a great deal about the size, age, and composition of our universe. But within our galaxy, the Milky Way, supernovas are exceedingly rare. How can you study something that cannot be duplicated in a laboratory, would fry you if you got close to it, and rarely even occurs?

That is where mathematical sciences enter the story, via computer simulation. In scores of applications, from physics to biology to chemistry to engineering, scientists use computer models—whose construction requires the formulation of mathematical and statistical models, the development of algorithms, and the creation of software—to study phenomena that are too big, too small, too fast, too slow, too rare, or too dangerous to study in a laboratory.

While scientists and engineers have long been able to write down equations to describe physical systems, before the computer age they could only solve the equations in certain highly simplified cases, literally using a pen and paper or chalk and a blackboard. For example, they might assume the solutions were symmetric, or simplify a problem to two or three variables, or operate at only one size scale or time scale.

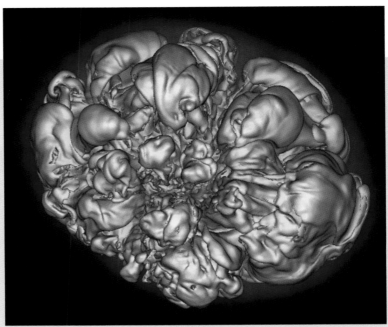

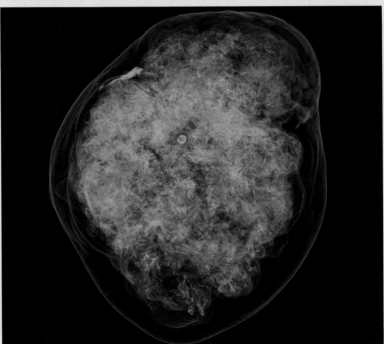

3 / *Image from a three-dimensional simulation of an exploding supernova. Reprinted with permission from Professor Adam Burrows, Princeton University.* /

Now, however, the scientific universe has changed. The study of supernovas is a perfect case in point. It is possible to create a rudimentary theory of supernovas by assuming that the star is perfectly symmetrical. Astrophysicists call this a one-dimensional theory because all of the quantities depend on one parameter, the distance from the center of the star. Unfortunately, it doesn't work: You can't get a one-dimensional star to explode, and so simulations based on that simplified model cannot represent all of the important aspects of this complex system. Of course, real stars are not so symmetric; they bulge at the equator, due to rotation. So astrophysicists began to simulate stars with a shape parameter as well as a size parameter, and they called these two-dimensional simulations. However, such simulations still cannot capture the behaviors of interest: Some fail to explode, while others explode but with less energy than a real supernova.

Only with fully three-dimensional simulations have astrophysicists started to produce supernovas with realistic energy outputs. And this tells us something important: The energy of the supernova must be coming from convection, a process that cannot be properly modeled in two dimensions. In a supernova, the core of a star collapses and then rebounds outward, forming an expanding shock wave. The shock wave then stalls as it runs into matter falling in from the outside the star. That's the hurdle that two-dimensional simulations have trouble getting over. But in three dimensions, the matter inside the shock wave starts to churn as it is irradiated by neutrinos, like soup being heated in a microwave oven. This convection reenergizes the shock wave over a period of several seconds, and the star's contents explode out into the universe (see Figure 3).

While many are aware of the amazing gains in raw speed from Moore's law—the approximate doubling of computer hardware capabilities every 2 years—successful simulation on this scale also depends to an equal degree on new algorithms that perform the needed computations. For example, the transition from two to three dimensions invariably increases (usually by an enormous factor) the difficulty of a problem, requiring mathematical advances in representing reality as well as problem solving. Three-dimensional simulations on this scale are possible only through a combination of massive computing power and smart mathematical algorithms. The transition from a two- to a three-dimensional model requires more than simply running the same code with more data points. Often, new mathematical representations must be incorporated to capture new phenomenology, and new comparisons against theory must be made to assess the validity of the resulting three-dimensional model. More generally, advances in mathematics and statistics and improved algorithms provide leapfrog advances in computational capabilities. Scholarly studies have estimated that at least half of the improvement in high-performance computing capabilities over the past 50 years can be traced to advances in mathematical sciences algorithms and numerical methods rather than to hardware developments alone.

> Scholarly studies have estimated that at least half of the improvement in high-performance computing capabilities over the past 50 years can be traced to advances in mathematical sciences algorithms and numerical methods rather than to hardware developments alone.

4 / *Anton, a special-purpose supercomputer, is capable of performing atomically detailed simulations of protein motions over periods 100 times longer than the longest such simulations previously reported. These simulations are now allowing the examination of biologically important processes that were previously inaccessible to both computational and experimental study. Printed with permission from D.E. Shaw Research.* /

The value of simulation is not limited to real-world problems of huge scale: It is just as useful for tiny problems such as understanding processes within our cells. Many of the cell's functions are carried out by proteins—large molecules that fold into a precise shape to accomplish a particular task. For example, the proteins in an ion channel, which regulates the flow of ions across a cell membrane, need to fold autonomously into a pore that will admit a potassium atom into the cell but not a sodium atom. A mistake at the subcellular level can have implications that affect the whole body. In cystic fibrosis the channels that are supposed to transport chlorine ions don't work correctly, possibly resulting in a buildup of fluid in the lungs; in certain kinds of heart arrhythmias, the potassium channels do not properly regulate the movement of potassium ions, which can interfere with the normal muscle contractions that create each heartbeat.

At present, nobody knows how to take the chemical formula for a protein and predict the shape it will fold into. The shape is determined by the forces between the many atoms within the protein and between those atoms and their surroundings.

5 / *Snapshots of the folding and unfolding of a protein, obtained from a simulation of unprecedented length performed on the special-purpose supercomputer Anton. Multiple transitions are observed between a disordered "unfolded state" (red and gray) and an ordered "folded" state (blue). Printed with permission from D.E. Shaw Research. /*

Calculating the net result of all those forces is a daunting computational challenge, but simulations are getting close to that goal. Recently a special-purpose supercomputer managed to simulate the motion of a relatively small protein called FiP35 over a period of 200 microseconds (one five-thousandth of a second), during which time it folded and unfolded 15 times (see Figures 4 and 5). Again, while part of this capability was made possible by the special-purpose hardware, it also depends on mathematical advances. First, the precise but computationally intractable force field must be replaced by a good approximation based on empirical data and on simpler systems, and mathematical analysis is necessary to characterize the adequacy of the approximation. Second, computational algorithms have been developed to speed the computation of the interactions between atoms. For the story of one such algorithm, see a later section, "Fast Multipole Method: A Long-Term Payoff."

These success stories illustrate the kinds of problems that scientists now routinely ask computer simulations to solve. For decades, biology had only two modes of research—in vivo (experiments with living organisms) and in vitro (experiments with chemicals in a test tube). Now, there is a third paradigm, in silico (experiments on a computer). And the results of this kind of experiment are taken just as seriously.

Nevertheless, simulations face major challenges, which will be the focus of ongoing research over the next 20 years.

First, real-world processes often require simulation over a wide range of scales, both in space and in time. For instance, the core collapse of a supernova takes milliseconds, while the crucial convection step takes place over a span of seconds and the aftermath of the explosion lasts for centuries. Spatially, the thermonuclear flame of the supernova varies from millimeters to hundreds of meters during the explosion.

In biology, the range of scales is just as daunting. Subcellular processes, like the opening and closing of ion channels, are linked to events at the scale of a cell. These effects cascade upward, affecting heart tissues, then the heart, and finally (in the event of a heart attack) the health of the whole body. Likewise, the timescales also span a vast range: microseconds for the folding of a protein, fractions of a second for the choreography of a single heartbeat, minutes for a heart attack, weeks or months for the body's recovery. It is very difficult to incorporate all these scales into a single mathematical model.

Related to the multiscale problem is the multiphysics problem. Often the types of models used at different scales are incompatible with one another. Events at a subcellular level are often chemical and random, influenced by the presence or absence of a few molecules. In the heart, these events translate into electrical currents and mechanical motions that are governed by differential equations, which are usually deterministic. Multiphysics can also characterize a single scale: The heart is simultaneously an electric circuit and a hydraulic pump. It's not easy to reconcile and simultaneously model those two identities. Progress in such cases often depends on a combination of insights from the domain science and the mathematical sciences.

Given the complexity of simulations, model validation also becomes an important challenge. First the modeler has to make sure that the individual parts of the program are working as expected; for a complex simulation, this can be very difficult. Then he or she will test it to see if it reproduces the behavior of simple real-world systems and matches existing data. Finally, the model will be used to make predictions about genuinely new phenomena. But there is no universal procedure for deciding when a model is good enough to use, so to some extent model validation is still more art than science.

Another major challenge for simulations in the near future has to do with hardware and software. It goes without saying that any scientist who does simulations would like more computing power. That is the main bottleneck in the supernova and protein-folding simulations. Three-dimensional simulations are just barely feasible today, but astrophysicists would really like to go up to six dimensions! That would allow more accurate simulation of the velocity as well as the location of each particle.

But raw computing power is not the only solution. At the cutting edge of research,

> At the cutting edge of research, the importance of new and better algorithms cannot be overstated.

the importance of new and better algorithms cannot be overstated. To put it simply, you can wait 2 years for Moore's law to hand you a computer that is twice as fast—or you can get the same speedup today by developing better algorithms.

Apparent advances in raw computation speed do not translate directly, and perhaps not even indirectly, to simulations that are faster or more accurate. Today's expectation is that the high-end computers of the future will have huge numbers of very fast "cores"—processing units operating individually at extremely high speed—but that communication between cores will be relatively slow. Hence, software written for computers with a single core (or a small number of cores) will not be efficient, and standard computations, such as those for linear algebra, will need serious reworking by mathematical and computer scientists.

Mathematical Sciences Inside..

The mathematical sciences help to predict the path and strength of a tsunami following an earthquake or other oceanic event (such as a massive landslide or volcano eruption). Mathematical models underpin tsunami warning systems by estimating where a tsunami will make landfall, how high the waves will be, and how fast the waves will be traveling. More fundamentally, the mathematical sciences help to map the topography of the ocean floor and infer large-scale wave behavior from independent ocean tide gauges that are irregularly spaced and can be hundreds of miles apart. This knowledge is behind emergency warnings and evacuations, which help to avoid potentially devastating consequences.

Numerical models are used to simulate the earthquake, transoceanic propagation, and inundation of dry land. To save time in the event of an emergency, these simulations are run for a variety of possible earthquake sizes and locations, and these scenarios are then combined with ocean tide readings as they become available. This figure shows the predicted sea level increase (in cm) resulting from the deadly 9.0 magnitude earthquake off the coast of Japan in March 2011.

Tsunamis

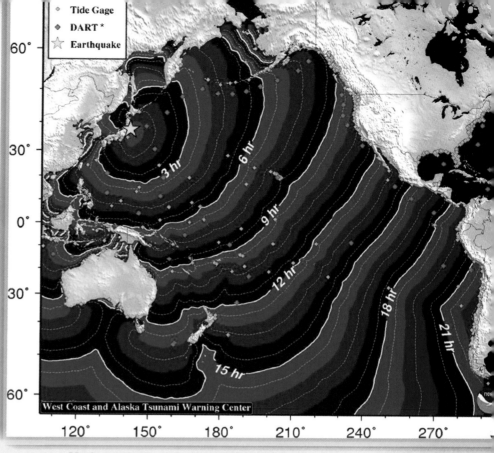

Mathematical science models help predict the timing and trajectory of tsunami waves based on ocean floor mappings and ocean tide gauge readings. This information is used to predict when tsunami waves will hit different coasts. *Deep Ocean Assessment and Reporting of Tsunamis

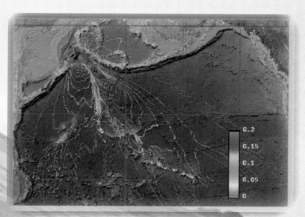

High-resolution computational models are used to simulate wave-heights for a traveling tsunami, shown here for two different earthquakes. These estimates help to identify evacuation zones and routes. The impacts of tsunamis vary widely, due to local topography, long-term sea level rise, annual climate variability, monthly tidal cycles, and short-term meteorological events.

Bayesian Inference

Not an Enigma Anymore

During World War II, the German army, navy, and air force transmitted thousands of messages using an encrypting machine called Enigma. Little did they realize that British mathematicians were eavesdropping on them. (In fact, the secret did not fully come out until the 1970s.) The battle of Enigma was, in its way, just as important as any military engagement. It was won in part by a statistical method called Bayesian inference, which allowed code breakers to determine probabilistically which settings of the Enigma machine (which changed daily) were more likely than others (see Figure 6).

In the years after the war, Bayesian analysis continued to enjoy remarkable success. From 1960 to 1978, NBC forecast the results of elections using similar techniques. The U.S. Navy used Bayesian analysis to search for a lost hydrogen bomb and a wrecked U.S. submarine. Yet these successes were not shared, for security or proprietary reasons. In academic circles Bayesian inference was rarely espoused, for both historical and philosophical reasons (as explained below).

Over the last 30 years, however, Bayesian analysis has become a central tool in statistics and science. Its primary advantage is that it answers the types of questions that scientists are most likely to ask, in a direct and intuitive

6 / *The World War II-era Enigma is one of the most well-known examples of cryptography machines and was used to encrypt and decrypt secret messages. Photo by Karsten Sperling.* /

way. And it may be the best technique for extracting information out of very large and heterogeneous databases.

For example, in medical applications, classical statistics treats all patients as if they were the same at some level. In fact, classical statistics is fundamentally based on the notion of repetition; if one cannot embed the situation in a series of like events, classical statistics cannot be applied. Bayesian statistics, on the other hand, can better deal with uniqueness and can use information about the whole population to make inferences about individuals. In an era of individualized medicine, Bayesian analysis may become the tool of choice.

Bayesian and classical statistics begin with different answers to the philosophical question "What is probability?" To a classical statistician, a probability is a frequency. To say that the probability of a coin landing heads up is 50 percent means that in many tosses of the coin, it will come up heads about half the time.

By contrast, a Bayesian views probability as a degree of belief. Thus, if you say that football team A has a 75 percent chance of beating football team B, you are expressing your degree of belief in that outcome. The football game will certainly not be played many times, so your statement makes no sense from the frequency perspective. But to a Bayesian, it makes perfect sense; it means that you are willing to give 3-1 odds in a bet on team A.

The key ingredient in Bayesian statistics is Bayes's rule (named after the Reverend Thomas Bayes, whose monograph on the subject was published posthumously in 1763). It is a simple formula that tells you how to assess new evidence. You start with a prior degree of belief in a hypothesis, which may be expressed as an odds ratio. Then you perform an experiment, or a number of them, which gives you new data. In the light of those data, your hypothesis may become either more or less likely. The change in odds is objective and quantifiable. Bayes's rule yields a likelihood ratio or "Bayes factor," which is multiplied by your prior odds (or degree of belief) to give you the new, posterior odds (see Figure 7 on page 22).

Classical statistics is good at providing answers to questions like this: If a certain drug is no better than a placebo, what is the probability that it will cure 10 percent more patients in a clinical trial just due to chance variation? Bayesian statistics answers the inverse question: If the drug cures 10 percent more patients in a clinical trial, what is the probability that it is better than a placebo?

Usually it is the latter probability that people really want to know. Yet classical statistics provides something called a p-value or a statistical significance level, neither of which is actually the probability that the drug is effective. Both of them relate to this probability only indirectly. In Bayesian statistics, however, you can directly compute the odds that the drug is better than a placebo.

So why did Bayesian analysis not become the norm? The main philosophical reason is that Bayes's rule requires as input a prior degree of belief in the hypothesis you are

> The main philosophical reason is that Bayes's rule requires as input a prior degree of belief in the hypothesis you are testing, before you conduct any experiments.

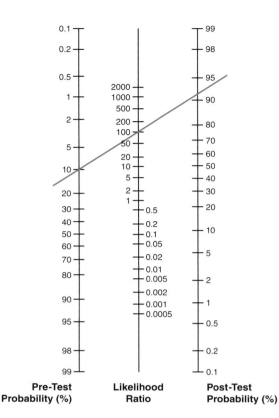

7 / Bayes's rule can be visualized using a nomogram, a graphical device that assists in calculations. Suppose, for example, a drug test is 100 times more likely (the "likelihood ratio") to give a true positive than a false positive. Also, you believe that 10 percent of the athletes participating in a particular sport (the "pre-test probability") use the drug. Now Athlete A tests positive. If you draw a line from 10 percent on the first scale through 100 on the second scale, it intersects the third scale of the nomogram at 91.7. This means that the "post-test probability" that Athlete A is using the drug is 91.7 percent. Adapted from the Center for Evidence Based Medicine, www.cebm.net. /

testing, before you conduct any experiments. To many statisticians and other scientists this seemed to be an unacceptable incursion of subjectivity into science. But in reality, statisticians have known ways of choosing "objective" priors almost since Bayes's rule was discovered. It is possible to do Bayesian statistics with subjective priors reflecting personal (or group) beliefs, but it is not necessary.

A second, and more practical, objection to Bayesian statistics was its computational difficulty. Classical statistics leads to a small number of well-understood, well-studied probability distributions. (The most important of these is the normal distribution, or bell-shaped curve, but there are others.) In Bayesian statistics you may start with a simple prior distribution, but in any sufficiently complicated real-world problem, the posterior probability distribution will be highly idiosyncratic—unique to that problem and that data set—and may in fact be impossible to compute directly.

However, two developments in the last 20 years have made the practical objections melt away. The more important one is theoretical. In the late 1980s, statisticians realized that a technique called Markov chain Monte Carlo provided a very efficient and general way to empirically sample a random distribution that is too complex to be expressed in a formula. Markov chain Monte Carlo had been developed in the 1950s by physicists who wanted to simulate random processes like the chain reactions of neutrons in a hydrogen bomb.

The second development for Bayesian statistics was the advent of reliable and fast numerical methods. Bayesian statistics, with its complicated posterior distributions, really had to wait for the computer age before Markov chain Monte Carlo could be practical.

The number of applications of Bayesian inference has been growing rapidly and will probably continue so over the next 20 years. For example, it is now widely used in astrophysics. Certain theories of cosmology contain fundamental parameters—the curvature of space, density of visible matter, density of dark matter, and dark energy—that are constrained by experiments. Bayesian inference can pin down these quantities in several different ways. If you subscribe to a particular model, you can work out the most likely parameter values given your prior belief. If you are not sure which model to believe, Bayes's rule allows you to compute odds ratios on which one is more likely. Finally, if you don't think the evidence is conclusive for any one model, you can average the probability distributions over all the candidate models and estimate the parameters that way.

Bayesian inference is also becoming popular in biology. For instance, genes in a cell interact in complicated networks called pathways. Using microarrays, biologists can see which pathways are active in a breast cancer cell. Many pathways are known already, but the databases are far from perfect. Bayesian inference gives biologists a way to move from prior hypotheses (this set of genes is likely to work together) to posterior ones (that set of genes is likely to be involved in breast cancer).

In economics, a Bayesian analysis of consumer surveys may allow companies to better predict the response to a new product offering. Bayesian methods can burrow into survey data and figure out what makes customers different (for instance, some like anchovies on their pizza, while others hate them).

Bayesian inference has proved to be effective in machine learning—for example, to teach spam filters to recognize junk e-mail. The probability distribution of all e-mail messages is so vast as to be unknowable; yet Bayesian inference can take the filter automatically from a prior state of not knowing anything about spam, to a posterior state where it recognizes that a message about "V1agra" is very likely to be spam.

While Bayesian inference has a variety of real-world applications, many of the advances in Bayesian statistics have depended and will depend on research that is not application-specific. Markov chain Monte Carlo, for example, arose out of a completely different field of science. One important area of research is the old problem of prior distributions. In many cases there is a unique prior distribution that allows an experimenter to avoid making an initial estimate of the values of the parameters that enter into a statistical model, while making full use of his or her knowledge of the geometry of the parameter space. For example, an experimenter might know that a parameter will be negative without knowing anything about the specific value of the parameter.

Basic research in these areas will complement the application-specific research on problems like finding breast cancer genes or building robots and will therefore ensure that Bayesian inference continues to find a wealth of new applications.

> **Bayesian inference is also becoming popular in biology. For instance, genes in a cell interact in complicated networks called pathways. Using microarrays, biologists can see which pathways are active in a breast cancer cell.**

Diffusion Tensor Imaging
A New View of the Brain

On battlefields and playing fields, from Iraq to Cowboys Stadium, one of the signature injuries of the past decade has been concussion. More than 300,000 soldiers suffered suspected concussions between 2001 and 2007. Nevertheless, it remains a difficult condition to diagnose because the damage to the brain is hard to see with conventional imaging techniques. The brain may look completely normal on magnetic resonance imaging (MRI) or on a computed axial tomography (CAT) scan, yet patients report ongoing effects, such as memory loss, headaches, sensitivity to light or noise, and depression.

A new imaging technique—a variant of MRI called diffusion tensor imaging—has revealed that the damage to concussed brains may lie not in the gray matter but in the white matter. For decades, neurologists considered the white matter (consisting of axons and glial cells) to be less important than the gray matter (which consists of neurons). They saw the white matter as a passive scaffolding for the brain's architecture. However, this view has changed dramatically in the last decade. If the brain is like a computer, then the gray matter can be compared to the processors, while the white matter can be compared to the communications grid that links those processors. Even the most powerful processors cannot work correctly if the pathways are destroyed or disrupted.

Besides concussion, a whole host of other brain functions and malfunctions are now linked to the white matter. Patients with schizophrenia, Alzheimer's disease, or deterioration due to a stroke, autism, and attention deficit disorder all have detectable changes in diffusion tensor imaging images of their white matter. Even during normal development and learning, the diffusion tensor imaging changes in intriguing ways.

FUELING
innovation and discovery

The revolution in our understanding of white matter—which has only just begun—would never have been possible without diffusion tensor imaging. And diffusion tensor imaging, in turn, would never have been possible without the mathematical sciences. The mathematics is hidden in plain sight: in that mysterious word "tensor" in diffusion tensor imaging. A tensor is a mathematical concept, developed in the 19th century, that generalizes the notion of vectors. Tensors have proved useful in a number of areas of physics.

To explain what a tensor has to do with white matter in the brain, it helps to start with how MRI works. An MRI machine (see Figure 8) creates a strong magnetic field, which causes the protons in the body to rotate and line up in a predictable way. Most of these protons are actually hydrogen atoms in water molecules; thus MRI is especially sensitive to the water (or fluids) in your body. It is an excellent complement to traditional x-rays, which see the dense, hard structures in your body but are relatively blind to the soft tissues. One of the most informative parts of the body to image with MRI is the brain, because it is squishy and it uses a lot of blood.

By modulating or pulsing the magnetic field in various ways, doctors can tune the MRI scan to detect different kinds of tissue in the body. In particular, one technique allows them to measure the displacement of water molecules over a short period of time—displacements that are due not to blood flow but to random jitters of the molecules, called Brownian motion. Because Brownian motion underlies the process of diffusion, this technique measures what is called the "apparent diffusion coefficient" in a tiny cubic region of the brain.

Already this imaging capability has led to fundamental insights about normal and abnormal brains.

8 / *Magnetic resonance imaging (MRI is an important medical imaging technique that allows internal structures to be visualized. Image courtesy of the National Institutes of Health Clinical Center, Center for Interventional Oncology.* /

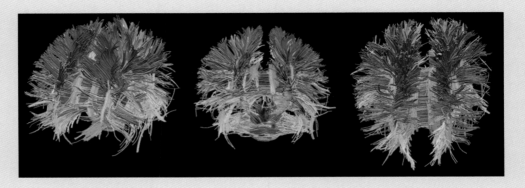

9 / Diffusion tensor imaging used to reconstruct network connections in the brain (tractography). Similarly oriented fibers are shown in the same color. Reprinted from Moriah E. Thomason and Paul M. Thompson, 2011, Diffusion imaging, white matter, and psychopathology, Annual Review of Clinical Psychology *7:63-85, with permission from Annual Reviews, Inc. /*

Beginning in the early 1990s, researchers noticed a puzzling fact: In the white matter, the apparent diffusion coefficient of a sample seemed to depend on its orientation with respect to the magnetic field. Tilt the sample and you would get a different diffusion coefficient. In 1991, a biomedical engineer had a eureka! moment: The dependence of the apparent diffusion coefficient on orientation wasn't a problem, it opened a path toward a solution.

This engineer knew something that most doctors didn't. In an anisotropic material—a material that is directionally dependent, such as a wood with a grain going in a particular direction or brain tissue that consists of layers or fibers—water doesn't diffuse equally rapidly in all directions. Water molecules move faster along the fibers and more slowly perpendicular to them. Over time, a tiny blob of water molecules will diffuse into an ellipsoid (or football) shape, with the long axis of the ellipsoid pointing along the fibers. The diffusion tensor contains all the mathematical information needed to graph this ellipsoid. It is not just a single number (like the apparent diffusion coefficient) but a 3 × 3 array of numbers. Starting with pork loins and working up to living human tissue, the experimental and mathematical procedures were developed for measuring the diffusion tensor from point to point within a sample and putting it into a three-dimensional image.

Diffusion tensor imaging was made to order for visualizing white matter, which consists mostly of axons, elongated cells that convey electrical impulses. Water diffuses rapidly along the length of an axon but slowly across the width. In addition, many but not all axons have a fatty sheath, called a myelin layer, which impedes the diffusion of water. (The myelin sheath is also what gives white matter its color.) Thus diffusion tensor imaging can both map out the direction of the brain's electric fibers (this is called "tractography" and is illustrated in Figure 9) and also detect the extent of myelination in various parts of the brain.

Already this imaging capability has led to fundamental insights about normal and abnormal brains. For example, biologists have known for a long time that the human brain starts out with little myelin, and that the axons gradually myelinate over childhood and adolescence. The myelination process seems to be associated with learning. With diffusion tensor imaging, researchers can now see this process in living humans. For example, they can see which parts of the brain are associated with reading and language acquisition. People with higher IQs in general tend to have longer, skinnier diffusion ellipsoids, suggesting greater fiber integrity or a greater amount of myelination. The fiber integrity (or "fractional anisotropy") seems to peak in the early 30s and gradually decreases thereafter; this may explain why memory and other cognitive processes decline gradually with age.

Likewise, diffusion tensor imaging points out areas of the white matter that are compromised in particular diseases. In schizophrenic patients, the fiber integrity is reduced in the part of the brain called the cingulate (responsible for error detection), the corpus callosum (responsible for communication between the brain hemispheres), and the frontal lobe. In autism, the deficits in fractional anisotropy occur in regions that are associated with processing social cues. Attention deficit hyperactivity disorder seems to be an exceptional case where the fractional anisotropy is too high rather than too low. And in concussion injuries, the fiber integrity is reduced near the site of the injury. This finding could be useful as both an objective criterion for diagnosis and a way of predicting which patients will suffer more serious long-term symptoms.

In the decade of the 2000s, research on diffusion tensor imaging took off, with the number of research papers doubling roughly every 2 years. Probably the most fundamental problem that remains is to distinguish when two fibers cross within a single cube (or "voxel," the three-dimensional analogue of a pixel) of the image. It has been estimated that as many as 30 percent of the voxels in a diffusion tensor imaging scan have more than one fiber passing through them. Unfortunately, the standard diffusion tensor cannot detect this fact. An ellipsoid has only one longest axis, and it cannot have two separate "bumps." If there are actually two fibers, diffusion tensor imaging will produce not two ellipsoids but a single, rounder ellipsoid. It will thus underestimate the fractional anisotropy in that voxel, and it may also draw the fiber pathways incorrectly.

One way to address the problem of crossing fibers would be to improve the resolution of the scans, so that each voxel is smaller. This would require MRI scanners with stronger magnetic fields—a trend that has continued throughout the past decade. But a less expensive alternative is to develop mathematical methods that would replace ellipsoids with more complicated diffusion surfaces. For example, a method called high angular resolution diffusion imaging (as shown in Figure 10 on page 28) combines magnetic resonance data with the principles of tomography, and it produces

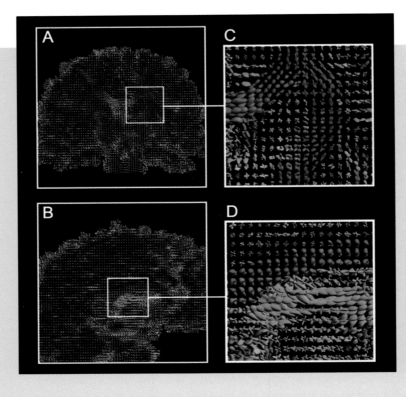

10 / High angular resolution diffusion imaging data used to show diffusion surfaces in three dimensions in the brain. This allows for a higher-resolution image than conventional diffusion tensor imaging but generates a great deal more data, making data mining and analysis more complex. Reprinted from Moriah E. Thomason and Paul M. Thompson, 2011, Diffusion imaging, white matter, and psychopathology, **Annual Review of Clinical Psychology 7:63-85**, with permission from Annual Reviews, Inc. /

spectacular detailed images of crossing fibers that would confuse an ordinary diffusion tensor imaging scan. However, it generates a great deal more data, necessitating advances in data mining and analysis. It is safe to say that much work remains to be done, from both the experimental and analytical sides.

Fast Multipole Method

A Long-Term Payoff

In school, we all learn that the problem precedes the solution. But in the mathematical sciences, it is sometimes the case that existing solutions suddenly become relevant to a new problem.

In the early 1990s, Vladimir Rokhlin of Yale University and Leslie Greengard of New York University had a solution. They had devised an algorithm called the Fast Multipole Method to speed up the solution of certain kinds of integral equations. Later, the American Institute of Physics and the IEEE Computer Society would name the Fast Multipole Method as one of the top ten algorithms of the century.

Louis Auslander, on the other hand, was a man with problems—lots of them. As the applied mathematics program manager at the Defense Advanced Research Projects Agency, his job was to match up defense-related problems with people who could solve them. When he heard about the Fast Multipole Method, he suspected that it could resolve an issue that had long bothered the Air Force, the problem of automatic target recognition.

The question is this: When you see an airplane on a radar screen, how can you tell what kind of airplane it is? Ideally, the radar should be able to recognize both friendly and enemy aircraft and determine which of the two kinds of plane it is. While this may appear to be a simple problem, it is in fact very difficult. Even if the plane is doing nothing to evade detection, every rivet on it, every bomb carried under its wings can change its radar profile. For an everyday analogue, compare a disco ball to a perfectly round, polished sphere. Even though their shapes are very similar, they reflect light very differently.

For years, the Air Force dreamed of having a library containing all the possible ways that a plane's radar signature could look. But to compile such a library, you would have to fly the plane past a radar detector thousands of times, from every possible angle and with every possible configuration of bombs or fuel tanks or other attachments. This would be prohibitively expensive and might be impossible for many enemy aircraft.

Alternatively, you could try to compute mathematically what the plane's radar signature should look like. If you could do that reliably, then it would be an easy matter to tweak the configuration to take into account bombs, fuel tanks, etc.

The problem of directly computing a radar reflection boils down to solving a system of differential equations called Maxwell's equations, which describe the way that electric and magnetic fields propagate through space. (A radar pulse is nothing more than an electromagnetic wave.) There was nothing new about the physics; the Maxwell equations have been known since the 19th century. The difficulty was all in the mathematics. Before the Fast Multipole Method, it took a prohibitively large number of calculations to compute the radar signature of something as complicated as an airplane. In fact, scientists could not compute the radar reflection of anything other than simple shapes.

It was known in principle that Maxwell's equations could be formulated as an integral equation, which is more tolerant of facets, corners, and discontinuities. To solve these, one must be able to calculate something called a Green's function, which treats the skin of the plane as if it were made up of many point emitters of radar waves, adding up the contribution of each source. This approach reduces Maxwell's equations from a three-dimensional problem to a two-dimensional one (the two dimensions of the plane's surface). However, that reduction is still not enough to make the computation feasible. Using Green's function to evaluate the signal produced by N pulses from N points on the target would seem to require N^2 computations. Such a strategy does not work well for large planes because the larger the plane is, the more data one needs to include and this approach requires an impractically large computational effort.

> ... the difference between computing the radar signature of a coarse approximation to an airplane and computing the radar signature of a particular model of aircraft.

The Fast Multipole Method builds on the insight that the problem becomes more manageable if the source points and target points are widely separated from one another. In that case, the radar waves produced by the sources can be approximated by a single "multipole" field. Although it still takes N computations to compute the multipole field the first time, after that you can reuse the same multipole function over and over. Thus, instead of doing 1 million computations 1 million times (a trillion computations), you do 1 million computations once, and then you do one computation 1 million times (2 million computations in total). Thus, Fast Multipole Method makes the more efficient Green's function approach computationally feasible.

A second ingenious idea behind the Fast Multipole Method is that it can be applied even when the source points and target points are not widely separated. You simply divide up space into a hierarchical arrangement of cubes. When sources and targets lie in adjacent cubes, you compute their interaction directly (not with a multipole expansion). That part of the Fast Multipole Method is slow. But the great majority of source-target pairs are not in adjacent cubes. Thus their contributions to the Green's function can be

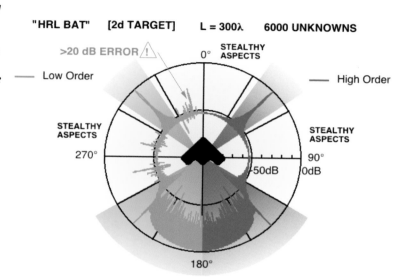

11 / Simulation of two-dimensional radar scattering from a stealthy airplane, similar in shape to the B-2 bomber. The front of the plane is at the top; two simulated radar signals are plotted in red and purple. On the left (red) is a computation using a low-order discretization. It incorrectly shows a considerable radar signal to the front and side of the airplane. On the right (purple) is a more accurate reconstruction of the radar signal, which would in practice be computed with the Fast Multipole Method. Note the near absence of a radar signal to the front and side of the airplane. (Actual three-dimensional data for the B-2 bomber are classified.) Reprinted with permission from Mark Stalzer, California Institute of Technology. /

computed quickly, using the multipole approximation. Because most of the calculation is accelerated and only a tiny part is slow, the overall effect is a great speedup.

In practice, the Fast Multipole Method has meant the difference between computing the radar signature of a coarse approximation to an airplane and computing the radar signature of a particular model of aircraft. While it would be tempting to say that it has saved the Air Force millions of dollars, it would be more accurate to say that it has enabled them to do something they could not previously do at any price (see Figure 11).

The applications of the Fast Multipole Method have not been limited to the military. In fact, its most important application from a business perspective is for the fabrication of computer chips and electronic components. Integrated circuits now pack 10 billion transistors into a few square centimeters, and this makes their electromagnetic behavior hard to predict. The electrons don't just go through the wires they are supposed to, as they would in a normal-sized circuit. A charge in one wire can induce a parasitic charge in other wires that are only a few microns away.

Predicting the actual behavior of a chip means solving Maxwell's equations, and the Fast Multipole Method has proved to be the perfect tool. For example, most cell phones now contain components that were tested with the Fast Multipole Method before they were ever manufactured.

At present the semiconductor companies use a slightly simpler version of the Fast Multipole Method than the original algorithm developed for Defense Advanced Research Projects Agency. The simpler version is applicable to static electric fields or

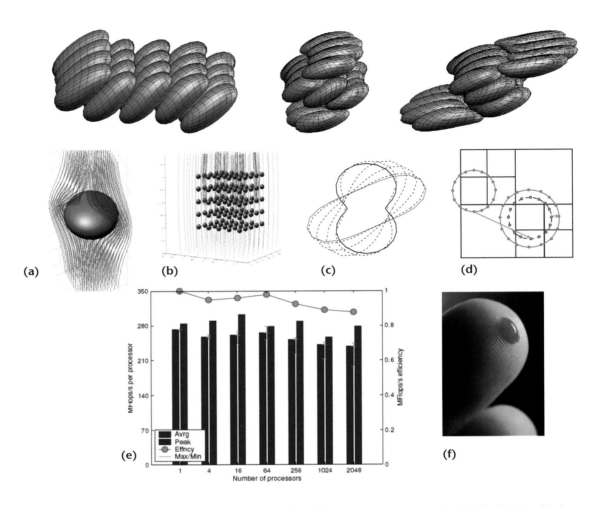

12 / *A simulation of red blood cells, performed on the Jaguar computer at Oak Ridge National Laboratory, won the Gordon Bell Prize for best use of supercomputing in 2010. The Fast Multipole Method was used to accelerate the computation of long-range interactions between red blood cells and plasma. A single-cell computation is shown in (a) and multicell interactions are shown in (b). Each cell's boundary is discretized (c). The hierarchical structure of the Fast Multipole Method computation is shown schematically in (d) and (e). The volume of blood simulated is shown in (f). This was 10,000 times the volume of any previous computer simulation of blood flow that accurately represented cellular interactions. Reprinted from A. Rahimian, I. Lashuk, S.K. Veerapaneni, C. Aparna, D. Malhotra, L. Moon, R. Sampath, A. Shringarpure, J. Vetter, R. Vuduc, D. Zorin, and G. Biros, 2010, Petascale direct numerical simulation of blood flow on 200K cores and heterogeneous architectures.* **ACM/IEEE Supercomputing (SCxy) Conference Series: 1-11, Figure 1 with permission from IEEE.** /

low-frequency electromagnetic waves. Believe it or not, even a 1-gigahertz chip—which operates at 1 billion cycles per second—operates at low frequency from the point of view of the Fast Multipole Method! That is because a 1-GHz electromagnetic wave still has a wavelength that is much longer than the width of a chip.

However, it is quite possible that the computer industry will one day graduate to optical chips, which will use light waves instead of electricity to communicate. A wavelength of light is much shorter than the width of a chip. So for an integrated optical chip, the high-frequency version of the Fast Multipole Method will become an essential part of the product development cycle. If so, the small investment in mathematical sciences that Auslander made in 1990 will pay off in an even bigger way three or four decades later.

Finally, it is worth noting that variants of the Fast Multipole Method are applicable to problems that have nothing to do with electromagnetism. The method is relevant to any situation where a large number of objects interact with one another, such as stars in a galaxy or red blood cells in an artery. Each red blood cell affects many nearby red blood cells, because they are packed quite densely inside a viscous fluid (the plasma). In addition, blood cells are squishy: they bend around curves or around each other. To take this into account, a good computer simulation needs to track dozens of points on the surface of each blood cell, and those points interact strongly with one another as the cell maintains its structural integrity.

In 2010, a simulation of blood flow based on the Fast Multipole Method won the prestigious Gordon Bell Prize for peak performance on a genuine (i.e., not a toy) problem of supercomputing (Figure 12). Researchers at the Georgia Institute of Technology and New York University used the Oak Ridge National Laboratory's Jaguar supercomputer to simulate the flow of 260 million deformable red blood cells (about the number from one finger prick). This smashed the previous record of only 14,000 cells and allowed the simulation to approximate the real fluid properties of blood. Although media attention focused on the supercomputer, the calculation would not have been possible without the Fast Multipole Method, implemented in a new way that was adapted to parallel computing. Ultimately, such simulations will help doctors understand blood clotting better and perhaps improve anticoagulation therapy for people with heart disease.

Mathematical Sciences Inside..

The mathematical sciences underpin many of the technologies on which national defense depends. Cutting-edge mathematics and statistics lie behind smart sensors and advanced control and communications. They are used throughout the research, development, engineering, and test and evaluation process. They are embedded in simulation systems for planning and for warfighter training. Since World War II, the mathematical sciences have been key contributors to national defense, and their utility continues to expand. This graphic illustrates some of those impacts.

The mathematical sciences are used in planning logistics, deployments, and scenario evaluations for complex operations.

Mathematical simulations allow predictions of the spread of smoke and chemical and biological agents in urban terrain.

Mathematics and statistics underpin tools for control and communications in tactical operations.

Mathematics is used to design advanced armor.

The Battlefield

Signal analysis and control theory are essential for drones.

Large-scale computational codes are used to design aircraft, simulate flight paths, and train personnel.

Signal processing facilitates communication capabilities.

Satellite-guided weapons utilize GPS for highly-precise targeting, while mathematical methods improve ballistics.

Mobile translation systems employ voice recognition software to reduce language barriers when human linguists are not available More generally, math-based simulations are used in mission and specialty training.

Modeling and simulation facilitates trade-off analysis during vehicle design, while statistics underpins test and evaluation.

Cellular Automata

Sublimely Complex

A sandpile grows in an hourglass, one grain at a time. Occasionally the impact of one sand grain will create a minicascade of grains. Even less often, a minicascade will trigger a macrocascade that changes the shape of the whole sand pile. How often do these cascades occur, and do they follow a predictable pattern?

A snowflake grows by attaching one molecule of water at a time. How does that process lead to the lacy, frilly shapes that we see in real snowflakes? What do we get if we grow the snowflake in different ways, using different attachment rules?

A city has two directions of traffic, east and north. At each intersection a car can proceed only if there is no other car already in the intersection. How many cars do you need to have in the city before gridlock ensues? Can the traffic self-organize into moving patterns?

These are three examples of a stylized mathematical model called a cellular automaton. Often inspired by real physical objects (such as sandpiles, snowflakes, and traffic), cellular automata demonstrate that amazingly complicated large-scale effects can arise from very simple local rules. They have been applied to other phenomena as diverse as avalanches and wildfires.

There is a widespread belief that highly complex systems can only be understood through computer simulation. However, mathematicians have found that simulation is far from the only way to discern patterns and phenomenology.

"Packard snowflakes," invented by Norman Packard in 1984, begin with a hexagonal lattice, like a honeycomb, in which one cell is filled by an ice crystal. One particularly interesting snowflake grows by the following rule: If an open cell is adjacent to 1, 4, 5, or 6 cells that are already filled, then at the next time step that cell fills and freezes. However, nodes that are adjacent to 0, 2, or 3 "frozen" nodes remain unfrozen at the next step.

After many steps, a shape emerges that bears a striking resemblance to a real snowflake (Figure 13). What if we let the Packard snowflake grow for as long as we want? Will all of the gaps in the snowflake eventually be filled?

In computer simulations, every gap seemed to fill eventually. But in 2006, it was shown that unfillable gaps do exist. The closest unfillable gap to the center is at least a billion cells away! Because a billion billion cells have to be filled before you can even get to this gap, it's no wonder that simulations never saw it. However, a billion billion is not an unrealistic number. It is less than the number of molecules in a milligram of ice. This example demonstrates the value of mathematical analysis and mathematical proof, which can provide insights that may be otherwise unavailable.

The cellular version of a sandpile starts with a rectangular grid and a distribution of sand grains on the grid that may be highly unstable. For instance, a thousand or a million grains could be stacked up into a tower (something quite unachievable with real sand). Like the Packard snowflake, the sandpile evolves by discrete steps. At each step in this simulation, a rule requires that any cell that is occupied by a stack of four or more grains of sand will redistribute one grain to each of its four neighbors. Of course, this may cause one of the neighbors to have four or more grains, so that the cell topples at the next time step. A cascade of toppling grains is thereby produced that can continue

13 / *A "snowfake" grows on a hexagonal grid by using simple attachment rules that model the adhesion of water molecules to a growing ice crystal. The resulting model (left, reprinted with permission from Janko Gravner, UC Davis, and David Griffeath, University of Wisconsin) looks remarkably similar to a real snowflake (right, reprinted with permission from Ken Libbrecht, California Institute of Technology.). The final macroscopic shape depends in very subtle ways on the local attachment rules on the molecular scale. /*

14 / *In a cellular automaton model of a sandpile, a billion grains of sand form complex patterns whose existence and shape can be confirmed from the mathematical theory of free-boundary problems. Colors denote the number of grains of sand in each cell (0, 1, 2, or 3). Reprinted with permission from David B. Wilson, Microsoft Research. /*

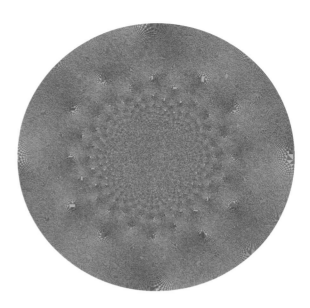

15 / *The rotor-router model is a deterministic version of a random process called internal diffusion-limited aggregation. The boundary has been proven to be circular; the "puckers" are still unexplained. Reprinted with permission from Tobias Friedrich (Max Planck Institute) and Lionel Levine (Cornell University). /*

for a very long time but ultimately settles down into a final configuration that looks like Figure 14. Note that in this figure, every cell contains 0, 1, 2, or 3 grains of sand and the cells are color-coded accordingly.

What is the shape of this final configuration? Does it depend on the number of grains of sand? What about the mysterious pattern of triangles that is so apparent in Figure 14?

In 2011, it was proved that at a large scale, the final shape is independent of the size. Thus a sandpile with a billion grains looks about the same as a sandpile with a million grains (only larger). Likewise, the pattern of colors stabilizes, so that every sufficiently large sandpile will look the same as the others (only larger or smaller).

The proof involves a mathematical model known as a free-boundary problem. Such problems arise, for example, in the study of combustion, glaciers, and stalactites, when a scientist is solving for not just an unknown function (say, the temperature in a moving flame) but also an unknown domain on which the function is defined. The theory of this particular free-boundary problem is less than 20 years old. Thus the sandpile problem, which might at first seem like an amusing game, is in fact intimately tied both to physics and to new developments in the mathematical sciences.

Another beautiful and conceptually important cellular automaton is called the "rotor-router." One version was designed to mimic internal diffusion-limited aggregation, which is more or less the flip side of snowflake growth. Imagine a crystal that instead of growing by attachment of particles from outside, grows by attachment of new particles from within. Each new particle wanders around the inside of the crystal at random until it hits a cell that has not been occupied yet and stops there.

Unlike the previous two examples, internal diffusion-limited aggregation involves randomness. A rotor-router model was designed to be essentially a derandomized version of internal diffusion-limited aggregation, where each new particle follows a prescribed sequence of turns that is designed to distribute its motion equally in each direction. In 2005, it was shown that the shape of a crystal grown by the deterministic rotor-router model is the same as the shape of one grown by internal diffusion-limited aggregation: a circle (see Figure 15). This result may appear specialized, but the principle it illustrates is much more general: Sometimes the average behavior of a random system can be well captured by a deterministic system. In such cases the randomness may not actually be an essential feature of the process.

The theory of cellular automata remains a very lively area of research, awash in examples, with many unexplored territories and relatively few guiding or unifying principles to join them. Cellular automata have been used to study avalanches, forest fires, landslides, and earthquakes, among others. The traffic model in Figure 16 shows that a traffic jam must happen before the city streets are 100 percent occupied. Simulations suggest that the onset of traffic jams happens when the density is between 30 and 40 percent, and no one yet has been able to close the gap between the theory and the experiments to really understand the dynamics of traffic. And there seems to be an intermediate and remarkably structured regime between freely moving streets and gridlock that could be described as a moving traffic jam whose existence has not yet been confirmed.

Though much remains unknown about cellular automata, it is exactly at such wild and untamed frontiers that the mathematical sciences grow. The connections between cellular automata and more classical mathematics, such as those mentioned above, bode well for the future development of the subject. These connections are like wires bringing electricity to the frontier.

> Though much remains unknown about cellular automata, it is exactly at such wild and untamed frontiers that the mathematical sciences grow.

16 / *Biham-Levine-Middleton traffic model (right, reprinted with permission from Alexander Holroyd, Microsoft Research):* In this idealized version of a grid of one-way streets, eastbound traffic (red) interacts with northbound traffic (blue). Jammed regions are solid, and flowing traffic forms dashed lines. /

Graph Spectra

Sparsest Cuts in Minimum Time

When chemists want to identify the molecules in an unknown sample—say, a rock from another planet—the first thing they do is measure its spectrum. That is, for instance, how space scientists detected water on the Moon and methane on Saturn's moon, Titan.

It may seem surprising that mathematical scientists are interested in measuring spectra too, in a whole host of applications that seem to have little to do with the original meaning of the word. To detect the edges of a cancerous region in a mammogram, they can use an "image segmentation" program that computes a spectrum. Similarly, problems like designing a computer chip that runs efficiently, identifying communities within a social network, or creating a computer network that won't shut down if some of the computers go offline, all involve analyzing the spectra of networks, or what mathematicians call "graphs."

To understand what a molecular spectrum is, think of a molecule as a collection of balls bound together with the aid of springs. The springs are always vibrating, and there are certain modes of vibration that, if left alone, will keep on going indefinitely. Each mode of vibration corresponds to a specific energy. The molecule prefers to absorb and emit photons with exactly the right energy to excite one particular mode. Those photons can be detected by a spectrometer, and the characteristic frequencies of the photons allow you to identify the molecule.

A network, which is a type of graph (a mathematical structure used to represent relationships between objects), is a more general mathematical version of the balls-and-springs model of a molecule. The most obvious difference, perhaps, is that networks of transistors or people don't vibrate and therefore don't have an obvious energy function associated with them. However, this difference is only superficial. In fact, mathematicians have derived a way of measuring energy on any network.

The way they do this is with something called a Laplacian, which has characteristic frequencies or eigenvalues, just like a vibrating molecule. Those frequencies collectively form a spectrum, and this spectrum is in many cases the best way to gain an overview of the network. For instance, it helps you find cliques (closely connected sets of nodes) and transportation bottlenecks.

But there is a problem of scale, because the networks of today are much larger than those in molecules. For example, the users of Facebook form a social network with nearly 1 billion active users. An Intel Xeon chip is a network with 800 million transistors. The Web is a network with 46 billion Web pages (and growing). Mathematical scientists have to design, in effect, a new kind of spectrometer to compute the spectra of such immense graphs.

Laplacians have a very long history—in fact, they predate both the concept of molecular spectra and the concept of graph theory. They were first defined in the 19th century to solve problems of heat flow and wave motion. The heat equation says that heat will flow toward any point that is colder than the average of its neighbors and away from any point that is warmer. The wave equation says that a point on a vibrating drumhead will accelerate in the direction of the average of its neighbors. In both cases, the operation "averaging the neighbors" is performed by a differential operator called the Laplacian (see Figure 17).

In the drum example, the eigenvalues of the Laplacian correspond to particularly simple modes of vibration, which we hear as the fundamental frequency and the overtones of the drum. In the heat equation, the eigenvalues have a slightly different interpretation. A cooling bar of metal has a relaxation time that governs how rapidly the bar cools off. This time is inversely related to the first (or smallest) eigenvalue of the Laplacian.

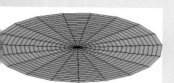 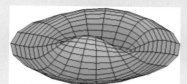 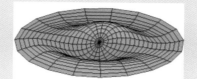 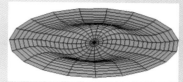

17 / Fundamental modes of vibration of a drum. Notice how certain frequencies of vibration isolate certain regions of the drum. Graph spectra are based on a similar idea, with the drumhead being replaced by a network of springs. /

In 1970, mathematician Jeff Cheeger proved that Laplacians are good at detecting bottlenecks in surfaces and higher dimensional objects called manifolds. The reason is especially apparent in the case of heat flow. If a metal bar is shaped like an hourglass, the constriction means that the first eigenvalue of the Laplacian will be small. In turn, this implies that the relaxation time will be longer than for a rectangular bar, and that is indeed what one measures.

The Laplacian on a network is defined analogously to the Laplacian on a surface. It is again computed by taking the difference between a function's values at a point and the average of its values at the neighboring points. In 1985, it was shown that Cheeger's result applied perfectly to this new setting. The first eigenvalue of the Laplacian, which is relatively easy to compute, is a good proxy for the width of the narrowest bottleneck, which is harder to determine.

A network has a bottleneck if it contains two (large) subgraphs with (relatively) few connections between them. (There are some subtleties connected with the words "large" and "relatively.") A driveway isn't a bottleneck, because if you cut it you take only one house out of the network. However, if two large cities are joined by only one bridge—such as the Ambassador Bridge between Detroit and Windsor, Canada—that bridge is definitely a bottleneck. Even if you include the Detroit-Windsor Tunnel, creating a second link between the cities, the ratio of the number of cuts required to disconnect the network (two) to the minimum number of people taken out of the network (216,000 on the Windsor side) is still very small (2/236,000, or 0.00001) (see Figure 18).

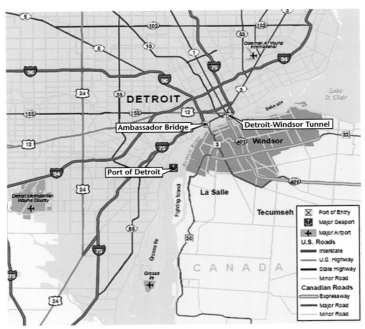

18 / The Ambassador Bridge connecting Detroit, Michigan, and Windsor, Ontario, is a very visible example of a "miminum cut." /

In a map of a city it is easy to spot the links that, if cut, have the largest effects on connectivity, especially if there is a river flowing through it. Such a link is called a sparsest cut, and in a typical graph, it is much harder to spot. But there are many instances in which we want to make those cuts, such as when we wish to decompose a network into simpler subgraphs with minimal effect on connections. For instance, a Laplacian-based method has been used to extract features from images. The surface of an object can be approximated by a mesh. The spectrum of this mesh can be used by a computer to identify distinct regions of the image without any human input.

Cheeger's work was important because it related something that was known to be very difficult to compute (sparsest cuts) to something that is much easier to compute (eigenvalues). It should be noted that Cheeger's theorem is an estimate, not a precise equation. However, the connection has proved robust enough to make the spectrum of the Laplacian an important tool in graph partitioning.

Spectral methods—a class of techniques to numerically solve dynamical (or constantly evolving) systems—can be used to design new networks as well as to study existing ones. For example, networks with the largest possible eigenvalues, called "expander graphs," have some remarkable properties. They convey information very rapidly, so it is easy to route messages from one point to another. They have no bottlenecks, so they cannot be disabled by the failure of a few nodes or links. Unfortunately, they are not so easy to design, so research in this area is currently a high priority.

Finally, although eigenvalues are easier to compute than sparsest cuts, even they get difficult when you are faced with monster networks consisting of millions or billions of nodes. Until recently, algorithms for finding the eigenvalues did not scale well and were not practical for such large networks. However, it now appears that this barrier has been breached.

In 2004, a method was devised for solving equations involving graph Laplacians that was significantly faster (in theory) than any previously known method. Like many great scientific advances, the strategy was wonderful, completely unexpected, and yet in retrospect completely natural. The idea is to replace the Laplacian of the graph with the Laplacian of a well-chosen subgraph. If this is done carefully, the spectrum is almost undisturbed yet the graph becomes much sparser. Imagine ripping out 95 percent of the streets in Manhattan and having everybody's commute time remain almost the same! That is roughly what this method managed to do.

There is still plenty of room for practical improvement, and some substantial improvements have already been made. The long-standing dream of finding sparsest cuts in approximately linear time—in other words, being able to understand the structure of a network almost as quickly as one can read in the data—now appears much closer.

Bioinformatics / Interpreting the Human Genome

On June 26, 2000, two biologists—Francis Collins of the international Human Genome Project and Craig Venter of Celera Genomics—stood side by side with President Clinton in the East Room of the White House and announced that they had finished sequencing the first draft of the human genome. All of a sudden, the molecular code that makes us human seemed like an open book. The genome was subsequently published in *Science* magazine.

Though it was heralded as a breakthrough in biology—and rightfully so—the cracking of the human genetic code also owed a great deal to the mathematical sciences. The Human Genome Project had begun in 1990 and was originally expected to take at least 15 years. However, around 1998 advances in the new discipline of bioinformatics—which combines biology with computer science, statistics, linear algebra, combinatorics, and even geometry—dramatically accelerated the project, turning it from a marathon into a 2-year sprint to the finish.

In the years since 2000, genetic sequencing has become even more dependent on mathematical science techniques. Next-generation sequencers have reduced the cost of reading an entire human genome from $300 million to $30,000 and the time from years to weeks. Further improvements, including the "$1,000 genome," are anticipated soon. The lower cost and quicker turnaround trend are continuing. The speed of information processing has now become the rate-limiting factor.

How did scientists assemble the human genome? The process is often compared to putting together a jigsaw puzzle. The analogy is a good one, but it is incomplete. In genomics, many of the pieces don't match, and some are duplicates. Also, many of the pieces come in pairs with a string glued to each piece, so you know roughly how far apart they are supposed to be in the puzzle. These complications present both opportunities and challenges for mathematical analysis.

Human DNA is a long molecule that is shaped like a spiral staircase, in which each step contains a pair of amino acids that fit together like a tongue and groove joint. Adenine (A) fits together with thymine (T), and cytosine (C) fits together with guanine (G). Each chemical fits only one of the others, so that the sequence of letters along one side of the staircase (GATTCC...) uniquely determines the corresponding sequence on the other side (CTAAGG...), which is conventionally read in the opposite direction (... GGAATC). Like a photographic negative, one strand is a template for duplicating the other (see Figures 19 and 20).

In all, human DNA contains about 3 billion "base pairs," or rungs of the staircase. The goal of the Human Genome Project was to list all of them in order. Unfortunately, chemists can sequence only a few hundred base pairs at a time. To sequence the whole genome, scientists had to chop it into millions of shorter pieces, sequence those pieces, and reassemble them.

The publicly funded Human Genome Project and the privately funded Celera Genomics adopted two different strategies, both of which eventually led to the same mathematical problem. You have millions of short (500-base) overlapping puzzle pieces that have been completely scrambled by the chopping process. There are enough pieces to cover the length of the genome seven or eight times over, so there are many overlaps between pieces. You want to use these overlaps as a guide to assemble the pieces into the longest possible sequence of contiguous regions.

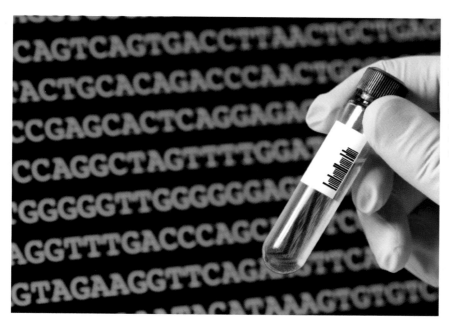

19 / Human DNA can be extracted from biological tissue such as skin and blood and a unique genetic sequence of amino acids can be determined. Image courtesy of the National Institutes of Health. /

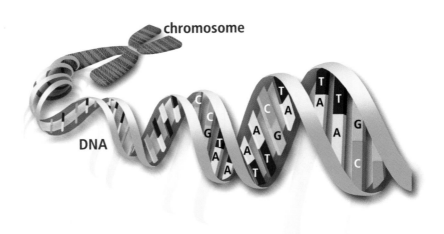

20 / *The structure of DNA: a double helix with matching base pairs of CG and AT. Image from U.S. Department of Energy Genomic Science Program.* /

If all of the sequence reads were perfectly accurate, matching the overlapping ones would be routine. However, about 1 percent of base pairs were misread, and this meant the overlapping jigsaw puzzle pieces would not match. The approach then became to find a good match rather than the best match (see Figure 21).

Another issue, somewhat more subtle, was the problem of repeats. Human genomes include many sequences that repeat identically in many places. These repeats were a big headache for genome sequencers because when a contiguous region ended with a pattern that occurred in many places, they had no idea which puzzle piece should come next.

The way around the problem, it turned out, was to take a longer snippet of DNA— say, several thousand base pairs long—and sequence both ends. Even though you can't sequence the middle, you can at least get a few hundred base pairs at each end and estimate how many base pairs lie between them. This gives you strings that link two jigsaw puzzle pieces together, including some that are on opposite sides of a gap or a repeat. These tethers create a scaffold to hang the "contigs" onto. Finally, the scaffolds could be wheeled into proper position by using the Human Genome Project's high-level map of the genome.

In the more than a decade since the completion of the human genome, the landscape has changed in at least two important ways. First, because a "reference" human genome is now available (in fact, many such reference genomes are), no human genome has to be sequenced from scratch. If you have a patient with cancer or with a genetic disease, you can zero in on the 0.1 percent of the genome that is different from the reference version and ignore the 99.9 percent that is the same. Thus the problem is not one of assembling the genome but of looking up sequences in the reference genome that are similar to (but slightly different from) your patient's.

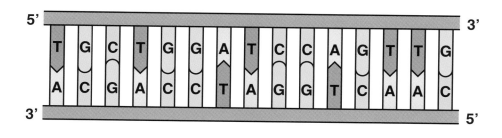

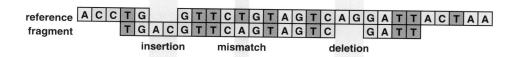

21 / The top image illustrates the sequence of C's, G's, T's, and A's in an untwisted segment of DNA. The bottom image shows typical errors in DNA sequencing: insertions, mismatches, and deletions. Reprinted with permission from the American Mathematical Society. /

Once again, the solution to this lookup problem came from far outside biology. In 1994, a "transform" was devised that dramatically speeds up the search for strings of text in a long file. Researchers created a table in which each row is a copy of the string, shifted either left or right. The rows are in alphabetical order. To search for a string like ATCTTG, you look for all the rows beginning with A, then for all the ones beginning with AT, and so on. Instead of searching through a linear string of 3 billion characters, you descend a tree with (in this case) six layers of branches. Once you get to the bottom, a transform identifies all the places where ATCTTG appears in the original string. Thanks to such indexing techniques, the reference genome can be searched in a fraction of a second.

A second transformation of genomics was the introduction of commercial next-generation gene sequencers, around 2004. Thanks to new advances in chemistry, biologists can now read hundreds of thousands of DNA snippets simultaneously. But the technology comes at a cost: The snippets have to be much shorter. One popular commercial sequencer can read fragments of only 50 to 75 base pairs, and another can manage only 100 to 150. The short reads are a double whammy for biologists. First, they need to collect much more data—typically the new machines will sequence enough snippets to cover the genome 30 times over. Second, a tiny 50-base read is more likely than a 500-base read to fall right into the middle of a repeated motif. The methods used by the first generation cannot deal with this increase in ambiguity.

Again, the right mathematics was already in existence and ready to be used, but it was unknown to biologists. The idea is to create a network in which nodes represent substrings of the genome and edges represent overlapping substrings. The first-generation methods amount to finding a path through a network that passes through each node once. This problem is known to take a hopelessly long time to solve. However, a better approach is to find a path that passes through each link of the network exactly once. (Note that it may visit some nodes repeatedly. These correspond to repeated strings in the genome.) This problem, called the Eulerian path problem, has a computationally efficient solution, which makes next-generation sequencing practical (especially for other animal species, which have no reference genome to consult).

The application of the mathematical sciences to the genome stands to have a great impact on society. A Battelle Memorial Institute study in 2011 concluded that the economic impact of the Human Genome Project has nearly reached $800 billion—quite a return on the U.S. government's $3 billion investment. And that doesn't even begin to account for the impact on humans, which is just beginning.

For example, in 2010 a 39-year-old patient was admitted to Barnes-Jewish Hospital in St. Louis with leukemia. There were two possible diagnoses with two different treatments. The patient had symptoms of a form of leukemia called APL, which responds well to chemotherapy. At the same time she had symptoms of another condition that would require a stem cell transplant, a risky procedure that can itself be fatal.

The standard genetic test for APL looks for two regions on chromosomes 15 and 17 that are swapped. That test came back negative. As an experiment, the patient's doctors spent $40,000 and six weeks to sequence her genome, comparing her skin cells to her cancer cells. In the cancer cells, they found about 77,000 base pairs of her chromosome 15 had been inserted into the chromosome 17—too small a fragment for the standard test to detect but easily detectable by genomics. Since then, they have identified similar genetic alterations in two other leukemia patients.

The doctors went on to treat the patient with the anti-APL drug instead of the dangerous stem cell transplant. It eliminated her cancer cells, and she was still in remission 15 months later. This is exactly the sort of individualized medicine that doctors dreamed the Human Genome Project would make possible. Though such medicine is still experimental today, with continued progress in bioinformatics it will probably be common a generation from now.

> The application of the mathematical sciences to the genome stands to have a great impact on society

Geometry and Physics / Endlessly Intertwined

"Philosophy is written in this grand book—I mean the universe—which stands continually open to our gaze, but it cannot be understood unless one first learns to comprehend the language in which it is written. It is written in the language of mathematics, and its characters are triangles, circles, and other geometric figures."

So wrote Galileo Galilei in 1623, at the dawn of the scientific era. Nearly four centuries later, we can only marvel at his prescience, because what he wrote then is even more true today. The secrets of the universe are still written in geometric terms, although the figures Galileo wrote about have now been replaced by more exotic and abstract ones: manifolds, fiber bundles, and Calabi-Yau spaces.

In the early 1800s, it was shown that the familiar Euclidean geometry, which has been taught since the ancient Greeks and is still taught in high schools today, is only one of an infinite variety of possible geometries. Euclidean geometry is flat—it is the geometry of a tabletop, infinitely extended. By contrast, non-Euclidean geometries are curved. They may have the positive curvature of a sphere, or they may have negative curvature, which is harder to visualize but may be compared to the frilly surface of some leafy vegetables.

In the 1850s, Bernhard Riemann took another bold step forward, describing spaces in which the curvature could change from point to point within the space. Riemann's geometry also allows space to take on any number of dimensions—two, three, or even more. He called these curved spaces "manifolds" (see Figure 22 on page 50).

For some time these new geometries remained just a mathematical curiosity. But in the early 1900s, Albert Einstein used Riemann's mathematics as a language to express his theory of general relativity—a theory in which gravity results from the curvature of four-

A long list of profound discoveries followed from the equations that Einstein wrote down in 1915: black holes, the expanding universe, the big bang, and dark energy.

 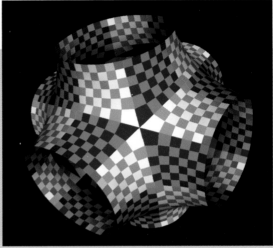

22 / *Two-dimensional manifolds are also known as surfaces. A sphere (left) is positively curved, while the surface on the right, which resembles a six-connection pipe fitting, is negatively curved. Reprinted with permission from Gerard Westendorp.* /

dimensional space-time. A long list of profound discoveries followed from the equations that Einstein wrote down in 1915: black holes, the expanding universe, the big bang, and dark energy. To understand any of these ideas fully, you have to learn Riemannian geometry. Somewhere, Galileo must be smiling.

But Einstein's general relativity was only the beginning. Similar geometric constructions underlie the field theories that describe particle physics. The discovery of antimatter, in 1932, grew directly out of an attempt to reconcile relativity with the quantum-mechanical description of the electron. The equations predicted extra solutions that seemed like positively charged electrons. We now call them positrons. They are the key ingredients in positron emission tomography, or PET scans, which are used to study the workings of the human brain.

In the later 1930s and 1940s, physicists and mathematicians started losing touch with one another. Physicists started thinking about fields that permeate all of space, which they called "gauge fields." (Examples include the electromagnetic field and the weak and strong nuclear forces.) Meanwhile mathematicians, for different reasons, became very interested in a new kind of geometric space, called a fiber bundle, which is roughly like a curved space with a quiver of arrows attached at every point (see Figure 23 for an example). It wasn't until the 1970s that mathematicians and physicists realized that they were doing the same thing. The physicists' gauge fields were like individual arrows in the mathematicians' quiver of arrows.

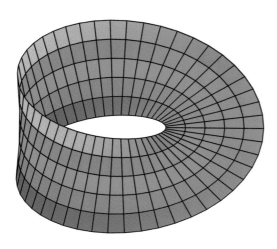

23 / *The Mobius band is the simplest nontrivial example of a fiber bundle. The fibers are shown in red. The twist given to the fibers makes the Mobius band topologically different from an ordinary cylindrical band.* /

24 / *A cross-section of a quintic hypersurface. Reprinted with permission from Paul Nylander, http://bugman123.com.* /

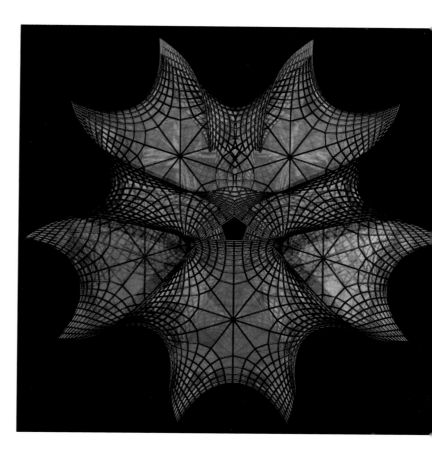

The cross-fertilization of ideas between the mathematical sciences and theoretical physics continues to this day. In the late 20th and early 21st centuries, string theory was formulated as an approach to unifying gravity and quantum physics into a theory of everything. Like all other theories in physics, it is highly mathematical—but the necessary mathematics has not yet been invented. There is still no rigorous context for the calculations that string theorists do, nor do mathematical scientists know the extent to which these techniques are valid.

However, the study of string theory has led to some important applications of the mathematical sciences. For example, since the 1800s, mathematicians have studied the solution sets of polynomial equations, such as a fifth degree polynomial in four variables. An example of such a polynomial is $x^5 + y^5 + z^5 + s^5 + t^5 = 0$, which is known as a "quintic hypersurface" (see Figure 24.) These surfaces contain figures that Galileo would have recognized.

And why did string theorists care about quintic hypersurfaces? Because string theory postulates that the universe has six extra, unseen dimensions that are curled up into a tight ball. Except that "ball" is not really the correct word. They actually form a manifold—a type of space discovered by mathematical scientists.

The list of interactions between geometry and physics could go on and on. It is difficult to speculate where it will lead next, but it is virtually certain that unexpected ideas for both disciplines will continue to grow out of the interaction. Galileo's words continue to hold true: Geometry is still the language spoken by the universe.

Probability and Statistical Physics

Connecting Microscopic and Macroscopic

In 1827, a botanist named Robert Brown noticed that grains of pollen suspended in water did a strange sort of dance when examined under a microscope. At first he thought the pollen was alive. But in 1905, Albert Einstein explained the real cause of "Brownian motion," which had nothing to do with biology. The grains are constantly buffeted by collisions with water molecules, which cause them to jiggle in random directions.

It is surprisingly common for random microscopic events to produce predictable effects at a macroscopic level. The Brownian motion of any single particle of smoke is highly unpredictable, yet smoke spreads in a room at a predictable rate. Iron atoms make innumerable random choices on which way to spin, but at a predictable temperature their spins spontaneously line up and the iron becomes magnetized. The voids in a porous material may be distributed randomly, but at a certain density, which is predictable, they connect up and the material becomes permeable. There are three kinds of "random path" here: the meanderings of a smoke particle, the boundary between spin-up and spin-down atoms in a magnet, and the path of water through a rock. But remarkably, all these disparate phenomena (or at least a simplified mathematical model of each one) can be brought under one roof.

In 2000, a universal mechanism was discovered that shows how microscopic disorder can lead to macroscopic order for two-dimensional systems. This discovery is now called Schramm-Loewner evolution, which allows precise calculations of macroscale phenomena that could until then only be predicted nonrigorously. Not only that, the mechanism applies to the random processes mentioned above as well as to others. A single parameter, κ (the Greek letter kappa), distinguishes Brownian motion ($\kappa = 8$)

25 / Schramm-Loewner evolution for different values of k. For example, critical percolation (top) gives trajectories with κ = 6. The middle image shows κ = 0.5, and the bottom image κ = 8/3. Reprinted with permission from Scott Sheffield, Massachusetts Institute of Technology. /

from measures of the boundaries between small regions of materials ("magnetic grains") that are aligned magnetically (κ = 3) or paths for water percolation (κ = 6) (Figure 25). Schramm-Loewner evolution is a wonderful unified description of all these separate phenomena, illuminating how disorder can create order.

The most essential feature of Schramm-Loewner evolution is a symmetry property called conformal invariance. Conformal invariance has two components: scale invariance and rotation invariance. The first means that a Brownian trajectory will look just the same at any level of magnification. If you blow it up by a factor of 10, it will look just as jiggly as before. The formerly small bounces will become big bounces—but you will see new, even smaller bounces that you couldn't make out before. Rotation invariance means that Brownian motion has no preferred direction. For example, in a closed room, smoke will go everywhere.

Consider a crack that grows randomly inward from the edge of an infinite pane of glass. According to the idea on which Schramm-Loewner evolution is based, the pane of glass can always be "healed" by a conformal transformation, deforming the glass in a way that pushes the crack back out to the boundary. If the crack grows, the glass can be healed again. In this never-ending process of cracking and healing, the attachment point of the crack moves around. In fact, it jiggles very erratically along the edge of the glass. Does this sound familiar? The attachment point is actually undergoing one-dimensional Brownian motion. The intensity of the jiggling is described by the parameter κ; larger values of κ correspond to more intense jiggling and to more jagged cracks. The basic idea of Schramm-Loewner evolution converts any two-dimensional, conformally invariant random process into a one-dimensional Brownian motion. Many questions become simpler after they are restated as a Schramm-Loewner evolution. What is the probability that the trajectories of two pollen grains in a petri dish will intersect before the grains reach the edge of the dish? What is the probability that water will percolate from one side of a

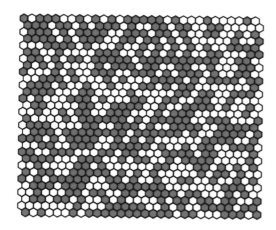

26 / Critical percolation model. When each cell has an equal probability of being red or white, the red and white cells connect up into very long networks, and the interface between them is a curve described by Schramm-Loewner evolution. Reprinted with permission from Michael Kozdron, University of Regina, Saskatchewan. /

rectangle to the other before it escapes out the top or bottom? What is the fractional dimension (or roughness) of the outside of a cloud of smoke? All of these questions have precise answers now.

Schramm-Loewner evolution is a purely mathematical construct. There is no known physical mechanism that can duplicate the cracking-and-healing process described above. It's "mathemagic" in the best sense.

Conformally invariant processes are particularly relevant to the physics of phase transitions, such as the freezing of water or the magnetization of iron. (Just as there is a freezing point for water, above a certain temperature, iron will not magnetize; below that temperature, it will.) These processes are scale-invariant because it is precisely at a phase transition that, when the temperature is dropping, small-scale, local correlations become large-scale as the ice crystals or iron crystals lock into place.

Though Schramm-Loewner evolution is a key to understanding such random processes in two dimensions, it has two caveats. First, it is anything but routine to prove that a given random process corresponds to a particular value of κ. There are some processes, such as the growth of polymers (which are like self-avoiding random walks), where the appropriate value of κ is strongly suspected but not rigorously established.

Second, Schramm-Loewner evolution is—unfortunately—limited to two dimensions. It seems very likely that three-dimensional random processes cannot be classified by a single parameter such as κ. It is likely also that the critical exponents describing the correlation of nearby molecules are not as simple as they are in the two-dimensional case. So 21st century mathematical scientists still have their work cut out for them as they try to explain phase transitions in our three-dimensional world.

However, Schramm-Loewner evolution provides a model for how a theory of phase transitions might look. Research based on Schramm-Loewner evolution has twice won the Fields Medal, one of the highest honors in the mathematical sciences. An award like this shows the remarkable amount of esteem among mathematical scientists for a discovery that is scarcely a decade old.

> So 21st century mathematical scientists still have their work cut out for them as they try to explain phase transitions in our three-dimensional world.

Mathematical Sciences Inside.

Remember how magical it seemed when CD players started to appear in automobiles? How could a precision instrument, which has to detect pits less than a micron across, possibly function in an environment where it is routinely jolted distances that are tens of thousands of times larger than that? The magic is not in the shock absorbers, it's in the mathematical sciences. A method called maximum likelihood sequence estimation, based on the statistical technique called maximum likelihood, works out the most likely sequence of 1s and 0s recorded on the disk and compensates for the noise and errors created by the bumpy car ride.

Many other technologies that we now take for granted are based on mathematical ideas. Other inventions that seem visionary today but might be commonplace 20 years from now, likewise depend on math. To illustrate that point, the table below lists 10 inventions whose patents cite a method from the mathematical sciences. The data are from the Google patent database at www.google.com/patents, accessed on August 3, 2011. The terms are ordered roughly by their frequency of occurrence in the Google database.

Q Fast Fourier Transform (FFT)

The industry standard way of decomposing an electronic signal into its constituent frequencies, based on samples taken at regular time intervals.

This patent is for a "monolithic" silicon chip that can compute FFT's; such chips are used in digital image processing, speech recognition, and transmission as in cell phones.

Patent No. 4547862 (1985) TRW, Inc.

Q Correlation coefficient

A basic statistical method for determining how closely related two paired sets of data are.

An optical scanner locates a "bull's-eye" on a label by finding the correlation between scanned pixel sequences and the expected sequence for a cross section of the bull's-eye.

Patent No. 6122310 (2000) United Parcel Service

Q Viterbi algorithm

Algorithm used in cell phones and CD and DVD players to decode noisy signals. Its key idea is to use "soft," probabilistic decision procedures.

This patent is one of hundreds that tweaks the original version, here by speeding up the "traceback" part of the algorithm.

Patent No. 6904105 (2005) Intel

nventions

Q Elliptic curve

Algebraic structure used in public-key cryptography—for example, to authenticate the user of a smart card.

In this patent, users can pick their own elliptic curve instead of selecting one from a centrally managed registry.

Patent No. 6446205 (2002) Citibank

Q Complex number

Numbers of the form $a + bi$ (where $i = \sqrt{-1}$).

Applications such as FFT require integrated circuits capable of adding and multiplying complex numbers, as described in this patent.

Patent No. 4858164 (1989) United Technologies

Q Minimal surface

Surfaces, such as soap films, that have the least area spanning a given boundary.

This patent proposes the Schwarz triply periodic minimal surface as a scaffold for regenerating human bone and organ tissue.

Patent No. 7718109 (2010) Mayo Foundation

Q Support vector machine (SVM)

Recently discovered (1995) method for partitioning data into classes.

SVMs are used in an implantable "brain pacemaker" for Parkinson's disease patients, to determine when the patient is having a seizure or movement disorder.

Patent No. 12/694035 (applied 2010) Medtronic, Inc.

Q B-spline

The industry standard method of representing smooth surfaces, used in computer-aided design and manufacturing.

B-splines have become popular in recent years with video game manufacturers; in this patent, they are used to generate smooth motions of three-dimensional figures under user control.

Patent No. 5982389 (1999) Microsoft

Q Quaternions

Hypercomplex numbers used primarily for composing spatial rotations.

This patent is for a toothbrush that will automatically track its location relative to the user's teeth. Quaternions are used to compensate for motion of the user's head.

Patent No. 12/866,381 (applied 2010) Philips

Q Conjugate gradient method

An iterative method for solving linear equations ($Ax = b$) or energy minimization problems involving many variables.

Used in this patent to compute the electronic structure of simple molecules like glass. The energy depends on thousands of variables, each representing a possible electron orbit.

Patent No. 6106562 (2000) Corning